Rambramha Sanyal

Hours with Nature

Rambramha Sanyal

Hours with Nature

ISBN/EAN: 9783337030636

Printed in Europe, USA, Canada, Australia, Japan

Cover: Foto ©berggeist007 / pixelio.de

More available books at **www.hansebooks.com**

HOURS WITH NATURE.

BY
RAMBRAMHA SÁNYÁL, c.m.z.s.

SUPERINTENDENT, ZOOLOGICAL GARDEN, CALCUTTA;
AUTHOR OF THE "HAND-BOOK ON THE MANAGEMENT OF
ANIMALS IN CAPTIVITY IN LOWER BENGAL."

"O Lord, how manifold are Thy works! In wisdom hast Thou made them all: the earth is full of thy riches."

Psalms.

S. K. LAHIRI AND CO.,
54, COLLEGE STREET.
1896.

PREFACE.

Should the following pages serve in any way to stimulate in our young men a love for the various interesting natural objects which surround us, my pleasant labours will not have been undertaken in vain.

I am greatly indebted to Professor D. D. Cunningham, C.I.E., F.R.S., and to Professor F. J. Rowe, M.A., of the Presidency College, Calcutta, for their valuable corrections and suggestions in the preparation of the book; in fact, the idea of writing these essays was first suggested by Professor Rowe.

ERRATA.

Page	Line	For	Read
19	5	Assam, and extending	and Assam, extending
22	2	rigions	regions
25	6	gardners	gardeners
26	9—10	insect-trap, for certain insects rendered	insect-trap for certain insects, rendered
30	7	Flara	Flora
31	16	Nathinel	Nathaniel
,,	17	Danisle	Danish
,,	28	distributed the	distributed to the
32	3	tittle	title
37	12—13	young, hopeful	young hopeful
42	1	naruralist	naturalist
56	24	indiscribable	indescribable
58	14	he mist	the mist
59	18	nigh	night
76	11	strike	shrike
91	4	childh	childhood, and yet
107	13	call them."	call them
122	2	they be able	to be able
125	15	content	contact
126	13	owing to this air is constantly enterning	owing to this, air is constantly entering
,,	15	salurated	saturated
127	9	open into	opens into
155	18	w, (or aw) are so	jaw are so
163	3	Bambay	Bombay

CONTENTS.

I.
AN EXCURSION.

NATURAL HISTORY OF THE DOLPHIN—OARSMAN'S STORY OF KINGFISHERS—HABITS OF KINGFISHERS—KING-FISHERS IN MYTHOLOGY—THE BOTANIC GARDEN—IN THE PALM GROVE—A SYLVAN BREAKFAST—A GROVE OF FOUR TREES—A CONGRESS OF BIRDS—THE BARBET—THE KOEL—THE MYNA—HOW SQUIRRELS ATE—NATURAL HISTORY OF THE SQUIRREL—THE ORCHID-HOUSE—THE FLOWER-GARDEN . . . . 1—27

II.
HISTORY OF THE BOTANIC GARDEN
(WITH SHORT BIOGRAPHICAL SKETCHES OF DISTINGUISHED BOTANISTS)

KYD—ROXBURGH—BUCHANAN—WALLICH—GRIFFITH—FALCONER—THOMSON—ANDERSON—CLARKE—KING—HOOKER—ROYLE—ROTTLER—A MORAL . . . . 28—36

III.
SHREWS AND MOLES.

THE GREY MUSK SHREW—THE MOLES—THE SHORT-TAILED MOLE—HEDGEHOGS . . . . 37—41

IV.
THE AQUARIUM.

ESTABLISHING AN AQUARIUM—STOCKING AN AQUARIUM—LIVE STOCK FOR THE AQUARIUM . . . . 42—47

V.

EXTRACTS FROM AN ANONYMOUS JOURNAL.

1. EARLY LESSONS IN ORNITHOLOGY—2. "IN MEMORIAM" JULY 18TH 18—.—IMAGES OF THE PAST—3. THOMAS PERIWINKLE—THE GREY PARROT—ANECDOTES OF GREY PARROTS—4. OF PARROTS IN GENERAL AND INDIAN PARROTS IN PARTICULAR—THE STORY OF THREE KAJLA PARROTS. . . . . . , ...   48—73

VI.

ROUND THE INDIAN MUSEUM.

BONELESS ANIMALS—POND-HUNTING—HALF AN HOUR WITH A MICROSCOPE—AMŒBA—PROTOZOA—MANY-CELLED ANIMALS—SPONGES, CORALS, &C.—WORM TO MAN—ECHINODERMATA—THE MOLLUSCA, OR SOFT-BODIED ANIMALS—OTHER FAMILIAR MOLLUSCA—THE COWRY SHELL . . . .   79—115

VII.

EXTRACTS FROM AN ANONYMOUS JOURNAL (II).

THE PARK—ON SOME TREES AND SHRUBS—ASHWATHWA—A LIVING BEING—HOW ROOTS AND STEMS CONVEY WATER—LEAVES HELP THE UPWARD FLOW OF WATER—HOW THE ASHWATHWA FEEDS AND DIGESTS.—SAL, TAL, TAMAL AND PIYAL—BAKUL—KADAMBA—A FINE AVENUE—BUTTERFLIES . . . . 115—139

VIII.

THE HOME FARM.—THE FARMSTOCK   ...   ... 140—148

IX.

INDIAN SNAKES.

THE RELATION A SNAKE BEARS TO OTHER ANIMALS—CHARACTERISTICS OF THE SNAKE—HOW IT MOVES IN

THE ABSENCE OF LIMBS—BODY COVERED WITH SCALES
—EYE, TONGUE AND TEETH—MOUTH AND TEETH OF
HARMLESS SNAKES—MOUTH AND TEETH OF A VENO-
MOUS SNAKE—VERY SMALL SNAKES—SOME LARGE
SNAKES—THE PYTHON IN CAPTIVITY—FALSE CHARGE
AGAINST SNAKES—THE COBRA—THE COBRA IN CAP-
TIVITY—THE SANKHACHUR—THE KARETA, RAJSAP,
CHANDRABORA AND CARPET VIPER—THE SEA SNAKE—
GENERAL REMARKS . . . . 149—165

X.

BY THE SIDE OF AN AQUARIUM.

THE KHALISA—THE TENGRAH—THE TITPUNTI—THE
MAGUR—THE SINGHEE . . . . , . . . 166—168

HOURS WITH NATURE

I.

AN EXCURSION.

ONE beautiful summer morning of the year 18—, we took boat at the Mullick ghát. It was yet very early, and the haze of twilight hung over the earth; but the breeze was fresh and cool, the morning fine, and the sky, the earth, and the water charming to look upon. As we glided down the river, a most lovely and picturesque sight came into view. To our left huge vessels laden with the wealth of a thousand shores lay serenely at anchor, a forest of trim spars, sharply defined against a morning sky of azure blue, stretched as far as the eye could reach; multitudes of country craft, large and small, were moored along the shore, round the buoys, and by the sides of the vessels. On the Howrah side, tall chimneys loomed in the distance, with a foreground of workshops, godowns, jetties, and cranes. There was a gentle ripple on the surface of the river, which softly lapped the sides of the boat. We were rowed further down, and the scene changed somewhat. Though still enchanting, it was no longer solemn, serene, and tranquil. Another day's toil and struggle had already begun, and signs of labour and man's pursuits were on every side. There was the fisherman's fragile dinghi paddling about, the steam launch ploughing the water below, and the kites and crows sailing in the air aloft. Presently, a few yards ahead of us and close to the starboard side of a vessel, we noticed a great commotion among these birds; on nearing the scene, we found that a bucketful of garbage and kitchen refuse had just been thrown overboard, and the excitement was due to the struggle among the crows, kites, and the gulls for the possession of a morsel apiece of this early breakfast. The crows were by far the most numerous and fussy, but while they spent much time in cawing and circling about, a kite would quietly swoop down, secure a bone almost before it had time to drop on the water and triumphantly carry it away to pick the flesh at leisure. We were in an observant mood, and did not fail to notice how the birds effected changes in the plane of their wings according as they wished to sail right or left, upwards or downwards; how the kite with its comparatively large wings could hover with ease, while the crow had to sustain its flight with much flapping. But what struck us as most curious and novel was the manner in which the crows secured the falling or floating bits of food with their claws, very much after the fashion of raptorial birds.

Our attention was suddenly diverted by a tumbling sound in the river close to our boat, and on looking towards the spot we just caught a glimpse of something greyish black in colour and smooth of skin disappearing in the water. It was a *Susu*—Gangetic Dolphin, (*Platenista gangetica* of Zoologists).

NATURAL HISTORY OF THE DOLPHIN.

The Dolphin has an elongated body, and a long compressed beak provided with large conical teeth, which are rather sharp-pointed in young, but become worn down as the animal grows old. Its dimensions vary from seven or eight to even twelve feet; the colour is uniformly black or greyish black. This aquatic animal is perfectly blind, because its eyes are not only rudimentary but completely buried beneath the thick, opaque hide of the head. There is, however, nothing strange in this deprivation of sight, which would be useless to a beast which lives in thick muddy water. Any one who has sailed often in the Ganges, the Brahmaputra, or the Indus cannot have failed to notice that these animals do not show themselves always and everywhere in these rivers. They are migratory in habits and desert such parts as become clear and shallow during the summer for others which are deep and dirty. Their food consists of fish and prawns, and also, it is said, crabs; but this requires confirmation. Dolphins are sometimes captured by fishermen, either accidentally or on purpose, as their oil is said to possess great efficacy in curing rheumatism and allied disorders.

We were now opposite the new docks at Khidderpore, and while looking at the operation of dredging going on in one of the graving docks, observed a bird perched on an over-hanging branch of a tree. It was sitting perfectly motionless and still, as if intently watching for something. Its head, face, the sides of the neck, body, and the upper part of the abdomen were chestnut brown; the chin, throat, middle of the neck, breast, and abdomen pure white; the beak, and feet were red.

It is a kingfisher. No, that cannot be, said another; the kingfisher has a black head, with white lines above the eyes; its back and wings are also black with white edgings; besides, it is a larger bird than the one sitting there.

Both of you are mistaken, broke in a third; the kingfisher is a common and familiar bird, frequenting rivers, tanks, *nullahs*, and water-courses. In fact, wherever there is water, be it in river or pond, or in a roadside ditch, there, one may be sure to find one or two of these tiny birds flitting about from branch to branch. It is a small bird, much smaller than the one sitting yonder.

Our discussion was interrupted by a merry peal of laughter suddenly ringing in our ears, and on looking about to ascertain its cause, we were rather disconcerted to find that it was at our expense. One of the oarsmen, a strapping lad of seventeen or thereabout, was attentively listening to our discourse about the identity of kingfishers. Being naturally intelligent and observant, he had improved his opportunities by taking mental notes of the habits and characteristics of birds and beasts which he chanced to meet in his roving life, and was conscious too of his knowledge; so that, our ignorance of such a common bird as the kingfisher had amused him vastly, but like the good boy that he was, he volunteered to tell us all about the bird.

OARSMAN'S STORY OF KINGFISHERS.

You must all know that riversides are favorite resorts of various kinds of birds and beasts. Then, there are others which though not living close by, habitually come to the bank at least once a day for a drink. I remember having once seen within twenty-four hours thirty-five different kinds of animals, including a tiger and a huge rhinoceros. This happened while we were pulling up a stream in the Sunderbuns. That, however, is another story. Let me tell you about the kingfishers that I have seen. There are several species of kingfishers, but my favorite among them is the Gureal. It is a very handsome bird with large stout bill of a bloodred colour. It is found along rivers and streams, but I have never seen it where there are no trees with dense foliage to hide it. It sits on a branch overhanging the water and watches for its food as you have just seen that one do. I dare say you have heard its peculiar call, a sort of *peer peer pur* uttered several times in succession. It is a powerful bird, I assure you, and quite a match for a hawk or a bird of that kind. I have heard that it breeds in hollow trees, but I don't know much about that. I have, however, once seen a nestling removed from the mud wall of a deserted shed. By the bye, some of you ought to find out whether they do breed in hollow trees or not. It would be adding something to the knowledge of the ways and habits of animals of our country.

This kingfisher has evidently a wide distribution. I have seen it about almost every river that I have frequented. The jungle-clad banks of the Brahmaputra and the Megna and the thick forests of the Sunderbuns appear to be its favorite haunts. While ferrying some Mughs across a creek in the Jessore Sunderbuns, I asked them if they knew that bird, meaning a Gureal which had just caught a crab. Yes, they were abundant in their country, they said. You know perhaps that Arrakan is the country of the Mughs.

The bird which you just saw and which formed the subject of discussion among you, is also a kingfisher and a near relation of the Gureal, although its beak is not so large as that of the latter bird. It is very common in Bengal and prefers to keep near human habitations rather than where forests are thick.

There is another kind of large-billed kingfisher which I have constantly seen in large numbers in the Sunderbuns. The beautiful sky-blue colour of the back contrasts very effectively with the brown and green of the rest of its plumage. It is not rare in Calcutta either, as I have often seen one or two flitting about in the trees near the Prinsep's ghát. Mughs tell me that it is also common in their country.

Did I not hear one of you gentlemen asserting that the kingfisher is a small bird? Yes, indeed, there are small kingfishers as well as large kingfishers. Whether you have noticed it or not, a minute ago one such flew across the bow of our dinghi. It is seen everywhere where there is water. I have often taken out young birds from deep holes in banks of rivers and reared them. One of them had grown so tame that I kept it loose. It remained perched on one of the poles of our little boat, and every now and then dived for fish and water insects. There are also other kinds of kingfishers; but we are now nearing the Botanic garden, and I must conclude the subject by telling you something about the

HABITS OF KINGFISHERS.

The food of kingfishers consists of small fishes, crustaceans, frogs, lizards, and aquatic insects. Patience, precision, and swiftness are essential to creatures which have to depend for their sustenance upon others equally wary, alert, and swift; and the kingfisher possesses these qualities in an uncommon degree. Swift as an arrow it darts on its prey as soon as one appears on the surface of the water, secures it in the twinkling of an eye, and then shoots back to its place of vantage. There is one species, the *Pied Kingfisher*, which, instead of watching for its prey from a fixed place, searches for it on the wing; it may often be seen hovering over a piece of water, and now and then darting down almost perpendicularly to secure a fish or a water-beetle. Kingfishers breed in spring and summer, and select some steep bank of a river or tank as their nesting place, and lay several round white eggs: they have been known to use fish bones as bedding for their nests.

KINGFISHERS IN MYTHOLOGY.

Kingfishers were well-known to the ancients, and stories half mythical, half superstitious, have gathered round these royal but familiar birds. It is said that Alcyone married Ceyx, who was drowned in a storm while going to consult the oracle. Alcyone being apprised in a dream of her husband's fate, and having discovered his body washed ashore on the morrow, threw herself into the sea, and was with her husband transformed into kingfishers—Halcyon. They were endowed with the virtue to keep the waters calm for a space of fourteen days while they built and sat on their nest on the surface of the sea. Hence the term "halcyon days," which has reference to this romantic legend of antiquity. Keats probably had this bird in view when he wrote:—

> "O magic sleep; O comfortable bird
> That broodest o'er the troubled sea of wind
> Till it is hush'd and smooth."

THE BOTANIC GARDEN.

The ancient and stately trees of the Botanic garden were now in view, and in another ten minutes we landed at the jetty lately erected for the convenience of visitors. Having ascended the ghát, we stood for a few minutes opposite the avenue of Oreodoxa palms in order to enjoy and contemplate the magnificent and picturesque beauty of the scenery which surrounded us, but, as we were eager to see the far famed Banyan tree of which we had heard so much and so often, we proceeded towards it without further loss of time, and in a short while were seated under the wide spreading branches of the glorious Banyan tree *(Ficus Bengalensis*, Linn.*)*, which looks more like a forest than a single tree. This forest-like appearance is caused by the aërial roots growing out of the branches and descending vertically to the ground to form supports for the horizontally spreading branches. This habit of sending down aërial roots is common in most of the trees belonging to the genus *Ficus*, but it has attained its greatest development in the banyan which sends down numberless branches to form

> " An ample shade,
> Cloistered with columned drooping stems and roofed
> With vaults of glittering green."

From the label fixed to the main trunk of the tree we gathered that it was about 51 feet in girth; that the aërial roots numbered 378, and that new roots were constantly forming. With regard to the origin of this tree there is a tradition current in the neighbouring villages to the effect that it began life as a parasite upon a date palm which it ultimately smothered. According to the same tradition a kind of sanctity is attached to the spot owing to the fact that a holy man (Fakir) used daily to sit under the date palm upon which the young banyan throve. Banyans, more often than not, begin life in this way. As a practical demonstration of the fact, there was a wild date with a young banyan thriving luxuriantly upon it within thirty yards of the spot where we were seated. There was yet another growing high and dry in a crevice formed by a crack in the coping of the wall in that part of the Garden. It is perhaps not generally known that certain birds are very fond of the fruits of these trees, and eat them greedily, and often drop the seeds in cracks of buildings and in the fissures of trees, where they vegetate. The banyan is a common tree in India, planted everywhere for its cool grateful shade. The flowers, which appear in summer, are enclosed in an excavated fleshy receptacle. Some of our common fruit trees, such as the *Dumar* (*Ficus carica*), *Kānṭhāl* (*Artocarpus integrifolia*), and *Tut* (*Morus alba*) belong to the same family as the banyan.

As other interesting and instructive objects awaited our visit, we dropped the subject of the banyan and proceeded towards that part of the Garden where the palmetum lay. On our way, and close to the Banyan tree, stood the monument of Roxburgh, the famous botanist, to whose long and loving labours the cause of Indian Botany is vastly indebted. It is a stone cylinder under a canopy, and bears an inscription.

IN THE PALM GROVE.

We were now in the midst of a lovely scene; the early sunshine of spring had thrown a cheerful glow over the shining ample foliage of the palms which grew luxuriantly on every side; a broad gravel path, fringed with ornamental shrubbery, ran across the middle of the grove, and here and there were placed comfortable garden seats for the convenience of weary visitors. There were numerous species of palms, of diverse shape and size, but we content ourselves with noticing such as are useful to us.

Sago Palm. (*Sagus farinifera*) It is a native of the Peninsula of Malacca, and the Malay Islands. Wallace, a great naturalist and traveller, describes it thus: "The sago tree is a palm, thicker and larger than the cocoanut tree, although rarely so tall, and having immense pinnate spiny leaves, which completely cover the trunk till it is many years old. It has a creeping root stem, like the Nipa palm, and when about ten or fifteen years of age sends up an immense terminal spike of flowers, after which the tree dies. It grows in swamps or in swampy hollows of the rocky slopes of hills, where it seems to thrive equally well as when exposed to the influx of salt or brackish water." The pith of the tree yields the sago of commerce, which is the staff of life to the inhabitants of many of the Malay Islands, and which is so useful to mankind in general. Its mode of preparation is as follows :—"When Sago is to be made, a full-grown tree is selected just before it is going to flower. It is cut down close to the ground, the leaves and leaf stalks cleared away, and broad strips of the bark taken off the upper side of the trunk. This exposes the pithy matter, which is of a rusty colour near the bottom of the tree, but, higher up, pure white, about as hard as a dry apple, but with woody fibres running through it about a quarter of an inch apart. This pith is cut or broken down into coarse powder by means of a tool constructed for the purpose—* * * *. By successive blows of this narrow strips of the pith are cut away and fall down into the cylinder formed by the bark. Proceeding steadily on, the whole trunk is cleared out, leaving a skin not more than half an inch in thickness. This material is carried away (in baskets made of the sheathing bases of the leaves) to the nearest water, where a washing machine is put up, which is composed almost entirely of the sago tree itself. The large sheathing bases of the leaves form the troughs, and the fibrous covering from the leaf stocks of the young cocoanut, the strainer. Water is poured on the mass of pith, which is kneaded and pressed against the strainer till the starch is all dissolved and has passed through, when the fibrous refuse is thrown away, and a fresh basketful put in its place. The water, charged with sago starch, passes on to a trough with a depression in the centre, where the sediment is deposited, the surplus water trickling off by a shallow outlet. When the trough is nearly full, the mass of starch, which has a slight reddish tinge, is made into cylinders of about thirty pounds weight, and neatly covered with leaves, and in this state is sold as raw sago."

The Indian Sago Palm *(Caryota urens)*, which resembles the sago palm in appearance, and is closely allied to it, is a native of the various mountainous parts of India, including tropical Sikhim.

This palm is also of great economic value to the inhabitants of the countries where it grows, as it also yields a kind of inferior sago from its pith.

The sweet Arenga. *(Arenga saccharifera)* is a denizen of the shores of the Malay Peninsula and Islands, Siam, and the Moluccas. This beautiful and stately palm has a straight, elegant, and columnar trunk, naked below, but above and near the base of the sheaths entirely covered with horse-hair-like fibres which issue in great abundance from the margins of the sheaths. In young trees the whole trunk is covered with these sheaths, which gradually drop off as the tree grows old. It repays cultivation by yielding sugar, sago, and excellent fibres for cables and cordage.

The Cocoanut palm. *(Cocos nucifera)* is a common and well-known tree, and requires no description. It is, however, not generally known that there are several species of cocoanuts, including those flourishing in the South Sea Islands, and the whole of the Brazilian coasts. Though it thrives inland some hundreds of miles from the coast, and sometimes at elevations varying from three to four thousand feet above the level of the sea, the sea-shore appears to be its chosen and congenial home. Owing to the great commercial value of its products, the tree is largely cultivated in almost all the Islands of the Malayan Archipelago, in Ceylon, the Laccadives, and on the Malabar and Coromandel Coasts. It is needless to say that it is also cultivated, though to a much less extent, in Bengal. It grows wild in distant, isolated, and uninhabited islands.

Date Palm.—Of the date palm we saw several species thriving in the Garden. The well-known *Khajur* (*Phœnix sylvestris*) though extremely common all over India is a tree of economic importance. It yields large quantities of juice, which is boiled down into sugar, or fermented for distillation. During the winter, the fresh juice is also drunk by the people. Mats and baskets are largely made from its leaves.

Mode of extracting Date palm Juice.—The lower leaves and their sheaths are removed, and a notch cut into the pith of the tree near the top; a small channel, made of a bit of palmyra leaf or of bamboo, is inserted into the notch; down this channel the juice trickles into an earthen pot suspended to receive it.

Its Cultivation.—Although generally growing wild in many parts of India, the date palm is not uncommonly cultivated in Southern India, Behar, and N. W. Provinces. "High ground is selected for the plantation and it is kept perfectly free from undergrowth, and the turf, from time to time, is ploughed up while the plants are maturing; the land can be cropped with oil-seed, and other dry crops. This lessens the original outlay, and the young palms are much benefited by the tillage between the rows. They are laid out in rows with twelve feet between each tree. Lime is a valuable manure for all saccharine plants and is usually added when it is deficient in the soil. If the trees are tapped before the end of seven years, they do not attain a full and healthy growth."

Phœnix farinifer—which is another kind of date palm, is a native of dry, barren, and sandy tracts on the east coast of Southern India. Its trunk yields a sago-like substance which is largely consumed by the poorer classes of the people during scarcity or famine.

The Arabian date palm—Of this, there were several varieties from Persia, Egypt, Bussorah, and the Persian Gulf. When carefully cultivated the Arabian date palm thrives in the climate and soil of India, especially in its drier regions. In countries where this species is largely cultivated, the method adopted for impregnating the female flower consists in making a slit in the spathe a little before it would burst naturally, thrusting into it a branch, or a part of the male spadix, and lightly tying them up with a string. The sweet date of commerce is a product of this palm.

The Hental palm.—A long row of *hental* (*Phoenix paludosa*) fringed the western border of the palmetum. It grows luxuriantly in the impenetrable thickets of the Sunderbuns, where, regardless of the tiger, the rhinoceros, and the deadly snakes which infest the jungles, the woodcutters penetrate, and collect loads of the smaller trunks of this palm, which are made into walking sticks. Tradition declares that snakes get out of the way of persons carrying *hental* sticks.

A SYLVAN BREAKFAST.

It was now breakfast time, and as the sun had mounted high and shone glaringly from a cloudless sky, we were in quest of a shady spot. Following one of the narrow gravel paths which branch out from the Casuarina avenue, we came upon a most picturesque part of the Garden. Here, there was no formal path, the one we were hitherto following having abruptly ended at the foot of a knoll on which stood a pretty arbour. Simple narrow footpaths on the green lawn were visible here and there, but they became indistinct and lost as we advanced. A high and extensive knoll overshadowed by a group of beautiful fig trees was chosen as our banqueting place. The scenery here was of remarkable artificial luxuriance and beauty. Around and in front of us the ground rose and sank and rolled along in undulations. Noble and stately trees covered the top and slopes of the mounds; a few of them, such as the *Poincianas* and others of the same habit, were now bare of leaves; the rest were clothed in beautiful summer foliage, and some looked gay and bright with blossoms. Far away and in front of us lay the serpentine lake, winding along the border of this artificial forest, and beyond it, a glimpse of the flower garden could be obtained through the colonnaded trunks of trees.

A grove of four trees may sound insignificant, but in reality it was a magnificent one. It consisted of four trees of a species of fig (*Ficus Comosa*), which grows abundantly in Southern and Central India. It has large spreading branches, with slender and pendant branchlets; and as these are much subdivided, a single tree overshadows a very large area. Owing partly to this, and partly also to the character of the leaves, which are smooth, oval, acutely pointed, and alternately arranged, it is one of the most elegant and beautiful of trees, and ought to be extensively planted in other parts of the country, where it is likely to thrive as well as it does in Bengal. The trees were all laden with an overflowing abundance of fruits which resembled *teparis*, but were of a deeper yellow colour. These attracted a large assemblage of birds of different species, and accounted for the bustle and commotion overhead in pleasing contrast with the general tranquility and repose which prevailed down below.

A congress of birds. There were the ubiquitous crows, fluttering, cawing, and asserting themselves; the fussy mynas chirping incessantly, and the cunning magpie gliding noiselessly among the branches. The beautiful green doves with bright yellow beak rose now and then fluttering their wings, and after wheeling about for a few seconds settled down again upon the same or a neighbouring tree. All of a sudden, a flock of mynas would leave the grove with a loud whirr, and betake themselves to another. So dense, however, was the foliage, that it was difficult to see all the birds, but there was no mistake about the presence of the *koel* and the *barbet*. Their respective call notes betrayed them.

The Barbet. (*Cyanops asiatica*, Lath). The Crimson-breasted barbet, or the Copper-smith bird of Europeans in India, has for its size, a remarkably loud call-note, which sounds like t-o-o-k, t-o-o-k, t-o-o-k, and which once commenced, it goes on uttering for some time, with a slight nod of the head at each call. It is asserted that the sound of the bird appears to come from a different direction to that from which it really proceeds. The bird is so common in India that any one with an inquisitive turn of mind can easily ascertain for himself whether there is any truth in the assertion.

The Koel (*Eudynamis honorata*, Linn.) the harbinger of spring, of "balmy breezes, and flowering meads," is a well-known bird in India. It is not usually gregarious in habit, though common pursuits sometimes bring several of them together. They and, for the matter of that, many other birds have more method and regularity in their habits than would appear from the roving life they lead. A koel may be observed to frequent the same group of trees for days and months, morning, evening, and noon, regularly about the same time singing from the same place. While we were speculating about their habits, a passing cuculine cry attracted our attention, and we saw a crow chasing a female koel with the utmost energy. The koel was probably detected by its pursuer in the act of depositing its eggs in her nest —a well-known habit among this species of birds.

The Myna. Of mynas, we noticed three species:—the common myna or *salik*, (*Acridotheris tristis*); the pied starling or *gosalik* (*Sturnopastor contra*); and the jungle myna (*Ætheopsar fuscus*). The two former are familiar birds, and common in every part of India. The jungle myna is not ordinarily seen near human habitations in a semi-domesticated state like its cousins, the *salik*, and *gosalik*. They are forest loving birds, and flocks of them may be found in the wooded regions of the Nilgiris, Mysore, the Ghâts, Central India, Nepal, Assam, and extending southwards to Burmah. They are not at all uncommon in Lower Bengal. The jungle myna is apt to be mistaken for a Bank myna or *gangsalik*, but a little observation will shew that it differs from the latter in having a small frontal crest at the base of the bill, and that it has an orange yellow bill instead of a red one with a yellow tip. These birds are somewhat migratory in habits within a limited area; and their migrations, like those of other animals, are determined by the abundance or scarcity of their food, which consists of fruits, berries and insects.

We were absorbed in observing the movements of the birds: how they jostled each other, and quarreled for what they might all have enjoyed in peace, and we did not notice what was going on around us. Places which looked perfectly deserted a quarter of an hour ago were now full of life. A large number of squirrels drawn from their sylvan refuges by the smell of food were moving about in all directions. They were very swift and shy; so that, if we looked at any one of them or moved a little, it darted away instantly and ascended the nearest tree, and whisked its tail and perked its ears, and stamped and chirped aloud and incessantly—

"With all the prettiness of feigned alarm,
And anger insignificantly fierce."

But owing partly to the liberal distribution of the remnants of our breakfast, and partly also to our attitude of perfect repose, they soon became more confiding, and approached nearer and nearer. It was very interesting to observe the nervous and suspicious manner of their movements. One would rush forward with lightning speed for a couple of yards, then stop short for a few seconds to gaze with its large prominent eyes, as if to satisfy itself that the coast was clear, and then would venture to make another dash.

How squirrels ate. On a bit of bread or a nut being thrown to one, it would spring upon it kitten-fashion, bound away for a few feet with the morsel in its mouth, and there sitting with its tail arched on its back, and holding the food in its forelimbs, begin devouring it bit by bit with its sharp chisel-like teeth. This attitude of sitting with its tail arched on its back suggested the ancient Greek name of the squirrel, which means an animal that sits under the shadow of its tail.

Natural history of the squirrel. The common striped squirrels are almost semi-domesticated in India and Ceylon, and always found near human habitations. They do not occur east of the Bay of Bengal; westward they extend as far as Sind and Beluchistan. There are about twenty different species of squirrels found in India and its dependencies, not to mention the numerous other kinds found in different parts of the world. But wherever found, and howsoever different they may be in structure and appearance, they have certain common characteristics. They all possess long and bushy tails, have five toes in the hind, and four toes in the forefeet with a rudimentary thumb. The character of their dentition is also peculiar: they have no canines or tearing teeth, but are provided with two large chisel-shaped and rootless front cutting teeth in each jaw. Between the incisors, as these front cutting teeth, are called, and the grinding teeth, there is a wide gap caused by the absence of the canines or tearing teeth ; there are ten grinding teeth in the upper, and eight in the lower jaw. The sharp chisel-shaped cutting teeth of a squirrel enable it, and for that matter all animals of the same order, to nibble at the hardest of nuts, fruits, roots, bark &c., which constitute their food. In captivity, they have been known to thrive well on gram, biscuits, cocoa-nut, and fruits.

THE ORCHID-HOUSE.

Our next object was the orchid-house. It is an extensive low structure, built of iron supports and galvanised wire netting ; the roof lightly thatched with ooloo grass *(Saccharum Thunbergi)* to protect the vegetation inside from sun and wind, and at the same time to let in enough of the former to vivify them. Graceful creepers of various species of Ipomœa formed a screen on all sides. Inside the house, the collection was rich and rare ; numerous species of ferns, begonias, anthuria, orchids, and hundreds of other kinds of plants from all regions of the tropics were displayed in luxuriant profusion, and well ordered confusion. Here for the first time we saw the wonderful pitcher plants of which we had read in books of travel, and which are said to attain their greatest development in the mountains of Borneo, Malacca, and Sumatra. There were two species of them exhibited:—Nepenthes Rafflesiana, and Nepenthes ampullacea, from Malacca and Singapore. The plants were not large, nor the leaves and pitchers numerous, but, such as they were, gave us a good id ea of what a pitcher plant is like.*

Numerous varieties of rare and beautiful ferns from different parts of the world throve in abundant profusion. We counted not less than sixty four species from such distant regions as Tropical America, Cuba, New Zealand, Australia, the South Sea Islands, the Malay Peninsula, the Philippine Islands, China, Mauritius, Bourbon, and various parts of India including Assam and the Himalayas. But the splendour of the display of orchids surpassed anything that we ever saw or imagined. Graceful clusters of Dendrobium from tropical Australia, the Philippines, the Malay Peninsula, and all parts of India hung on every side amidst a profusion of numerous other kinds; the delicate yellow colour of certain kinds

---

* The Pitcher plants belong to the order *Nepenthaciae* of Botanists. The plants are generally herbaceous with alternate leaves having pitcher-like pouches hanging from them. The apex of the leaf is prolonged into a tendril-like process, which afterwards expands and folds itself open to form the pitcher. Every pitcher is provided with a lid. Colour green, variously tinted and mottled.

contrasting beautifully with the soft mottled white Vanda and butterfly-like Phalænopsis.

Yet another scene of glorious beauty awaited our inspection and we hastened to

THE FLOWER GARDEN

which, though the season had somewhat advanced, was still a blaze of colours,—pink, white, yellow, blue, and crimson, all arranged in beds of appropriate design, and in such artistic fashion as to bring about a delightful effect. Apart, however, from its resplendent loveliness, redolent of sweet but subdued fragrance, the place had another charm for us. Here was a profusion of flowers mostly exotic, about many of which we had read in books of poetry and fiction; but, though they were familiar by name, we were as ignorant of their identity as babes unborn. The foremost flower to attract our attention was the Daisy, which recalled to our minds Wordsworth's beautiful lines:—

> Daisy ! again I talk to thee,
>   For thou art worthy,
> Thou unassuming Common-place
> Of Nature, with that homely face,
> And yet with something of a grace,
>   Which Love makes for thee !

The Daisy is indeed a flower of unassuming grace. The more we looked at it the more we realized the truth of the sentiments embodied in the following stanza :—

> I see thee glittering from afar—
> And then thou art a pretty star;
> Not quite so fair as many are
>   In heaven above thee !

> Yet like a star, with glittering crest,
> Self-poised in air thou seem'st to rest ;—
> May peace come never to his nest,
> Who shall reprove thee !

Centuries before Wordsworth, Chaucer had watched the Daisy, passing whole days leaning on his elbow and his side, and singing in the following strain :—

> " For nothing ellis, so I shall not lie,
> But for to lokin upon the Daisie,
> The emprise and flower of flowers all."

The same author gives us the origin of the name—

> "One called eye of the daie
> The daisie, a floware white and rede,
> And in French called La bel Margarete."

The Daisy is not the single and simple flower that it looks, but a cluster of little flowers or (florets) as they are botanically called, arranged upon a common receptacle and held together and surrounded by a cup-shaped green case. This green cup or case takes the place of a calyx or outer envelope of a flower and is formed of a number of little bracts or leaves that grow round the base of the small flowers. The outer florets are white and have strap-shaped corollas : the central florets are delicate yellow and have tubular corollas. The former have no stamens, but only a style with two stigmas, whereas the latter are quite perfect in that they have both stamens and pistils. The common Indian flowers গাঁদা (Tagetes patula) and চন্দ্রমল্লিকা (Chrysanthemum) are also compound flowers like the Daisy, and belong to the same Natural order. Although considered indigenous to the country, Tagetes (গাঁদা) is in reality a foreign plant, as it originally came from Mexico. But that is a story of olden days!

THE PANSY. In brilliant contrast with the modest grace of the Daisy and embosomed in their midst lay dominating over them in glory and pride a group of PANSIES or the *Butterfly flowers* of the Indian gardners.

It is a well known British flower. Among the various names by which it is designated in different parts of England and Scotland the "Heart's Ease" seems to be the most appropriate. The Heart's Ease of Indian poetry is however quite another plant. It is the famous Asoka or (sorrowless,) of which more hereafter. The Pansy is probably the flower which, as critics think, bewitched the Queen of the Fairies in the "Midsummer Night's Dream."

Pansies belong to the *Violet* family, and are all herbs, with stipulate leaves. The flowers are irregular with a spurred corolla; two of the stamens have honey-secreting apparatus: the pistil is connected with a one-celled ovary, with three rows of ovules. As a result of cultivation the garden flowers have generally some of the stamens and even the pistil transformed into petals. The coloration is superb, purple, black, white, yellow or crimson. The family is represented in India by a number of species, the common among them being the *Viola primulifolia*, Willd, which grows wild in various parts of Bengal, and blossoms more or less the whole year.

The SWEET VIOLET, from which the family derives its name, is an uninteresting flower to look at, but owing to its exquisitely delicate fragrance, it is a great favorite with everybody, and holds an important place among the cultivated garden plants.

Among other noticeable flowers those of *Snap-dragon* Anterrhinum (A. majus) interested us vastly owing to the peculiar arrangement of the corolla for the purpose of trapping insects. The flowers are of a purplish red colour or variegated with white. The corolla which is an inch long opens like a mouth when pressed between the finger and thumb and forms an efficient insect-trap, for certain insects, rendered attractive to the little creatures by the honey deposited within the tube. To have a taste of this sweet nectar the insects enter the treacherous tube. Once within it, all their efforts to find an egress prove unavailing.

Snap-dragon belongs to the *Secrophulariaceæ* or Mask-flowered family, and though only found as a cultivated garden plant in India, it grows wild on chalk cliffs and on old walls in the south of England.

In India plants of the same family are tolerably abundant. Our common *Kookshima (Celsea coromandeliana*, Vahl*)* is a familiar example. It grows wild in various parts of India, appearing generally as a weed in gardens or cultivated lands during the dry season. It is an erect, ramous, and downy herb; the lower leaves lyrate, the upper ones sessile cordate: flowers small and of a delicate yellow colour.

There was the PINK or DIANTHUS in its variegated beauty of white, pink, and scarlet. It holds an important place among the cold season flowers in every Indian garden. Its original home was China, but now it is a citizen of the world.

A group of PORTULACA is a glorious object during the middle of the day when the sun shines brilliantly and there are no bees about. It is a coy nymph which jealously guards the nectar stored within it. Its sensitive stamens begin to quiver and the corollas close upon them as soon as any bees appear on the scene.

Our common *Luniya** (*Portulaca Meridiana:* Willd) belongs to the same family as the *Portulaca* of the flower gardens. It is a small creeping annual with hairy joints, small, oblong, and fleshy leaves. Common in gardens, growing as weeds by the sides of metalled roads. The flowers open at noon and shut by 3 p. m., hence the specific name "Meridiana."

There were numerous other interesting kinds of flowers, all true emblems of purity, unsullied products of Nature, and fit objects to offer to Nature's God.

---

* লুন কিম্বা লূন ।

II.

HISTORY OF THE BOTANIC GARDEN.*

(WITH SHORT BIOGRAPHICAL SKETCHES OF SOME DISTINGUISHED BOTANISTS).

The founding of the Botanic Garden in Calcutta was the beneficent act of a noble mind. Colonel *Robert Kyd* of the Honorable Company's Engineers was an ardent horticulturist, and had gathered together in his private garden at Shalimar, a large collection of exotic plants. Deeply sensible of the benefit of an institution which might be made a source of botanical information for the possession of the Company, and a centre to which exotic plants of economic interest could be imported for experimental purposes, Colonel Kyd suggested the desirability of forming a Botanic Garden in Calcutta. His suggestions having been adopted by the Honorable Court of Directors, and practical effect having been given to it by the Government of India, he was appropriately appointed the first Superintendent of the Botanical Garden, which was founded at his suggestion. The earliest efforts of Colonel Kyd were directed towards the introduction of the trees which yield nutmegs, cloves, cinnamon, and pepper vines. It was however soon proved that the climate of Bengal is quite unsuited to these tropical species. The

---

*Compiled from the "Guide to the Royal Botanic Garden Shibpore." by Dr. George King, L.L.D., F.R.S., C.I.E.

equatorial fruits, such as mangosteen, langoat, dukko, and bread-fruit, as well as the temperate fruits of Europe were also tried with a similar result. It was thus demonstrated by practical experiment that certain natural products, many of them of a most desirable kind, cannot be grown in Bengal. Colonel Kyd also began the experiment of cultivating the teak tree, for the sake of its timber, then so invaluable for ship-building. But it became clear after an experience extending over a period of thirty-five years that, although the tree to all outward appearance grows well on the alluvial soil of the delta of the Ganges, its stems early become hollow near the base, and therefore useless for yielding timber of sound quality. Colonel Kyd continued to perform the duties of Superintendent of the garden until his death in 1793.

On his death *Dr. William Roxburgh*, the Company's Botanist in Madras was transferred from that Presidency, and installed as Superintendent of the Botanic Garden, Calcutta. For many years prior to his transfer to Calcutta, Dr. Roxburgh was engaged in studying the then little known Flora of the Northern Circars in the Madras Presidency. He was a most ardent and enthusiastic botanist, and a good gardener. He was the first botanist who attempted to draw up a systematic account of the plants of India. During his busy life in this country he prepared an account of Indian plants which contained systematic descriptions of all the indigenous plants known to him, as well as of many exotics then in cultivation in the Botanic garden and in the neighbourhood of Calcutta. Dr. Roxburgh continued to be Superintendent of the Garden until 1813, when he was obliged to proceed to the Cape on account of ill health. From the Cape he went to St. Helena, and from thence to Scotland; where he died in 1815. He took the manuscript of *Flora Indica* when he left India, intending to publish it during his residence in Scotland. His death prevented the execution of this plan. In 1820, Dr. Wallich and Carey printed, with some addition and interpolation, the first volume in two parts. In 1832, the remainder of the work was printed exactly as the author had left it by his sons Captains James and Bruce Roxburgh. It is an admirable production : the descriptions are accurate and graphic, and its authorship justly entitles Roxburgh to his title of the " Father of Indian Botany." It embraces descriptions and notices of most of the Indian plants from the stateliest trees to the lowliest herbs, and until the year 1872, when the publication of the " Flora of British India " was begun by the distinguished botanist Sir Joseph Hooker, Roxburgh's was the only single book through which a knowledge of Indian plants could be acquired. With a view to place the work within the reach of the poorest student, Mr. C. B. Clarke issued in 1874, a second edition of this excellent work at a nominal price of Rs. 5. So far, however, as the Indian students are concerned, his pious intention, it is greatly to be deplored, remains yet unrealised, as few, if any, have seriously taken up the study

HISTORY OF THE BOTANIC GARDEN.

of this branch of science. Besides the *Flora Indica*, Roxburgh published in three large volumes his *Plantæ Coromandelianæ*, being descriptions, with figures, of three hundred of the most striking plants of the Coromandel Coast.

Dr. Francis Buchanan (afterwards Buchanan-Hamilton) succeeded Dr. Roxburgh as Superintendent of the Botanic Garden. Dr. Buchanan-Hamilton was an accomplished botanist and zoologist. He travelled extensively in India, and collected materials for a gazetteer, published "An account of the agriculture of Mysore, in three volumes, quarto;" "An account of the Fishes of the Ganges," and other works. He was born in Perthsire in 1762, and died in 1837. He held charge of the garden for only a short time.

Dr. Nathniel Wallich lately Surgeon to the Danish settlement at Serampore succeeded Dr. Buchanan-Hamilton in the Superintendentship of the garden. Dr. Wallich was an accomplished and most energetic botanist, who during the earlier part of his term of office organised collecting expeditions into the remote and then little known regions of Kumaon, Nepal, Sylhet, Tenasserim, Penang, and Singapore. Dr. Wallich in fact undertook a botanical survey of the large part of the Indian Empire, and accumulated a large collection of materials in the shape of dried specimens of plants. These were taken to England and after being named by himself and other botanists were distributed to the leading botanical institutions of Europe. Dr. Wallich published, through the munificence of the Honorable Company, a botanical work of great merit under the title *Plantæ Asiaticæ Rariores*, in three volumes, illustrated by colored plates of a high degree of excellence. Dr. Wallich was a native of Denmark. After having been Superintendent of the Botanic Garden for thirty years, he retired in 1846 and died in 1854.

Dr. *William Griffith* filled the post of the Superintendent of the Garden during Wallich's residence in Europe for the purpose of naming and distributing his collection of plants. His premature death which took place in Malacca in 1845, deprived botanical science of one of its ablest and most industrious votaries. He was born at Ham Common, Surrey. He came out to India as a surgeon in the Honorable Company's Madras Establishment, and was one of the most brilliant of Indian Botanists. He accompanied the punitive expedition to Cabul in 1839, and formed one of the Botanical deputation which explored Assam in connection with the search for, and discovery of the tea plant; explored the Malayan Peninsula; amassed an enormous collection of dried plants. His collected works were posthumously published in nine volumes under the editorship of Dr. McLelland.

Dr. *Hugh Falconer* succeeded Dr. Griffith as Superintendent of the garden. Dr. Falconer was a palæontologist, well-known for his researches on the Sivalik Fossil Mammalia. He retired in 1855,.

Dr. *Thomas Thomson, F. R. S.,* who succeeded

Dr. Falconer, was a traveller and botanist of much ability, and the coadjutor of Sir Joseph Hooker in the collection and distribution of an extensive and well-known herbarium of East Indian plants; and he was also the joint author of the first volume of a new *Flora Indica*.

Dr. Thomas Anderson, a native of Edinburgh, a member of the Indian Medical service, and a botanist of distinction, held the post of the Superintendent of the Garden from 1861 to 1869. Dr. Anderson was the first conservator of Forests in Bengal, and the introducer of the quinine-yielding cinchona cultivation in the Sikhim Himalayas. This industry has been carried to such a successful issue in the plantation and factory under the wise and able direction of the present Superintendent of the Garden, that the Government hospitals and dispensaries have for years been supplied from this source with all the quinine required for them. He contributed to the journal of the Linnean Society many valuable papers on the difficult family of Acanthaceæ, and wrote for Sir Joseph Hooker's *Flora of British India*.

Mr. C. B. Clarke, F. R. S., acted as Superintendent of the Garden for two years subsequent to Dr. Anderson's departure from India, and during his incumbency he began the series of botanical publications which have earned for him so high a scientific reputation. Mr. C. B. Clarke was for sometime a member of the Bengal educational service. For his child-like simplicity, great scholarship, and devotion to science he was held in much esteem by every educated Bengali. His is an honored name even to this day.

Dr. George King, C. I. E., LL. D., F. R. S., the present Superintendent of the Garden, assumed charge of his office in 1871. Great changes have been effected in the Garden during these last twenty five years. The grounds have been laid out for landscape effects ; winding sheets of ornamental water have been formed, and pretty undulations have been thrown up. New roads and foot-paths have been laid, and bridges, herbaria and conservatories built. In fact, the Garden has improved in every department. As a botanist Dr. George King has earned a well-deserved distinction by the publication of a series of botanical works of great merit.

A host of other distinguished botanists have worked at Indian botany without being directly connected with the Botanic Garden, Calcutta. Foremost among them is Sir Joseph Dalton Hooker, K. C. S. I., C. B., F. R. S., the greatest living systematic botanist. Sir Joseph Hooker was for many years Director of the Royal Gardens, Kew, and was President of the Royal Society from 1874 to 1879. He travelled extensively in the Himalayas, and underwent many privations for the sake of collecting rare Himalayan plants, and of increasing his knowledge of Indian botany : he was taken prisoner by the Rajah of Sikhim. He is the author of two splendidly illustrated folio volumes on the Rhododendrons and other plants of the Sikhim Himalayas ; and also of the Himalayan Journal in two volumes. He is the editior and chief author of the *Flora of British India;* and joint author, with the late Mr. G. Bentham, of the most important English Botanical work of the century, *viz.* the *Genera Plantarum.*

John Forbes Royle, M. D., F. R. S., of the Honorable East India Company's Medical service, was another distinguished botanist. He was Superintendent of the Botanic Garden at Saharanpore from 1823 to 1831. His '*Botany of the Himalays,*' and *Fibrous Plants of India,* are works of great merit.

Rottler's name should not be omitted from this list. He was one of the group of scientific botanists who, during the latter years of the last and the earlier years of the present century, formed a Society for promoting the knowledge of Indian botany. Other members of this Society were Sir William Jones, Fleming, Hunter, Anderson, Berry, John, Roxburgh, Heine, Klein, and Buchanan-Hamilton.

A MORAL.

For us Indians, the history of the Botanic Garden, Calcutta, and the short biographical sketches of the distinguished botanists directly or indirectly connected with that institution, have a moral which may be expressed in the words of Gilbert White of Selborne recorded upwards of a century ago. " The productions of vegetation have had a vast influence on the commerce of nations, and have been the great promoters of navigation, as may be seen in the articles of sugar, tea, tobacco, ginseng, betel, paper &c. As every climate has its peculiar produce, our natural wants bring mutual intercourse; so that by means of trade each distant part is supplied with the growth of every latitude. But, without the knowledge of plants and their culture, we must have been content with our hips and haws, without enjoying the delicate fruits of India and the salutiferous drugs of Peru." We have many advantages, of soil, climate, vegetable and mineral productions; but powerless to use them for our good we choose to be "content with hips and haws," while the nation to which Gilbert White belonged has, since his days, rapidly grown in wealth, power and influence. The Tea plant must have been awaiting search and discovery in the inaccessible jungles of Assam for ages past; yet, it waited in vain until a Griffith with the magic power of knowledge made the jungle yield up its secret. Another botanist, as we have already seen, introduced the cultivation of Cinchona in India, which is now a source of revenue to the Government. Instances may he mutiplied to show that the science of Botany has a much wider scope of usefulness than that of collecting, naming, and classifying plants. It has influenced the trade and commerce of the world.

III.
SHREWS AND MOLES.

"A mole! a mole!" shrieked a little boy—as a musk-shrew crossed with lightning speed the floor of the low damp room where they—he and his private tutor—sat one stuffy July evening of the year 1874. Mahendra, for such was the name of the pupil, was an intelligent, bright-eyed, amiable lad of nine. The tutor's duty was to give him lessons at home for an hour or two every evening. Partly from a sense of duty, but greatly forsooth to make a display of his superior knowledge, as he had but recently learnt that a shrew was not a mole, the tutor asked his pupil where he could ever discover a mole in that place. "Why, Sir," readily answered young hopeful, with an intelligent and knowing look, "that animal that darted out of the room just now *was* a mole. I know it was: no other animal but a *Chhuncha* would make such a quick, sharp, squeaking noise, or leave such a disagreeable smell." "It was a *Chhuncha* to be sure", rejoined the tutor, "but not a mole". The pupil was surprised! His preceptor tried to explain to him that a *Chhuncha* is erroneously called a *mole*; it is in reality a *musk shrew*, so called from the strong musk-like odour which it emits. Puzzled, but not convinced, Mahendra immediately fetched an old manuscript vocabulary, and after having rummaged over its well thumbed pages for a few seconds triumphantly pointed to the word "Mole" with "*Chhuncha*" in Bengali opposite it.

This happened twenty years ago. As, however, our knowledge of Natural History does not appear to have much advanced, it may not be amiss to give here a brief account of moles and shrews. *The shrews* are small animals. The body is covered with soft hair; the head is long, and the pointed snout projects beyond the lower lip. All the animals of this family have small eyes. They are often popularly confounded with rats, but a moment's observation will show that the latter have no pointed nose or rounded ears resembling the human ear in shape. The form, number, and arrangement of teeth in shrews greatly differ from those in rats or other animals of the rodent family. Students interested in seeing and learning the difference of dentition in rats and shrews will do well to visit some museum, where skeletons of various kinds of animals are arranged in cases for the inspection and instruction of visitors.

The shrews have a wide range of distribution, being found in the temperate and tropical regions of Europe, Asia, Africa, and North America.

There are various species of shrews of which THE GREY MUSK SHREW [*Crocidura cærulea*—(Kerr)] is common all over India. It is the *Musk shrew* of the Anglo-Indians, *Chhuncha* of the Bengalis, and *chhuchandar* of the up-country Hindustanis; and it enjoys several other local names. That the animal is held in bad repute in India is evident from the fact that

Chhuncha or Chhuchandar is a term of contempt when applied to man.

Besides India, it is found in Ceylon and Burmah. Like all other animals of the same family it is nocturnal, and comes out at night to hunt about houses for insects, especially cockroaches which it devours greedily. It now and then utters, especially if alarmed, a sort of sharp squeaking cry such as arrested the attention of the little boy of whom mention has been made before. The disagreeable odour of a musk shrew is well known. On each side of its body is a gland which secretes a kind of evil-smelling fluid to which the odour of the animal is due. This bad odour is very useful to the shrew, as it serves to protect it from the attacks of such animals as live by preying upon others.

The food of the shrews consists of insects, to which they are very partial; but they appear to eat meat with equal relish. Some zoologists say that they have been known to attack frogs and eat scorpions.

THE MOLES are more perfectly fitted for an underground life than any other known animals. The fore-limbs are very broad and flat and furnished with large claws so as to render them capable of burrowing with ease. They are easily distinguished from the shrews by their thick rounded bodies, enormous forelimbs, and much shorter legs; the fur is peculiarly soft and velvety. The eyes are very small, in fact reduced to a point and generally covered by skin.* The short ears are concealed by the fur; the tail is very short.

The moles are found in Europe, portions of Asia, including Japan, and North America. In India they are confined to the Himalayas, and the hills of Assam and Cachar.

THE SHORT-TAILED MOLE. *(Talpa micrura*—Hodg.*)* is found in Nepal, Sikhim, and the hills south of Assam. It is not uncommon about Darjeeling, especially where there are decayed stumps of forest trees abounding in earthworms and larvæ of insects which afford food for the mole. The moles breed in summer, or at the beginning of rains, and have generally half a dozen young ones which quickly attain their full size.

While speaking of these animals it will not be uninteresting to make mention of another group of insectivorous animals known as :

HEDGEHOGS.

They are small rat-like insect-eating animals, of which the characteristic peculiarity is the presence of small spines on the back and sides. They are popularly called *Kantachuha or Shahichuha*—that is, spiny rat or porcupine rat. There are several species of hedgehogs found in India, of which Hardwicke's Hedgehog (*Erinaceus collaris*—Gray and Hardw.) inhabits North-Western India, the Punjab, and Sindh. The South Indian Hedgehog (*Erinaceus micropus*—Blyth) is an in-

---

* Except in one species, the Common Mole (Talpa europæa).

habitant of the plains of southern India. Hedgehogs, like moles and shrews, are nocturnal animals, and during the day remain concealed in the crevices of trees and rocks. They have the peculiar habit of completely rolling themselves up into a ball when touched or disturbed. Hedgehogs, like porcupines, have the power of suddenly jerking backwards, no doubt for the purpose of hurting their assailants. In captivity they have been known to feed upon grasshoppers, cockroaches, boiled eggs, and even bread and milk. It is also believed that they eat ants and millepedes.

IV.
THE AQUARIUM.

"No time" says a distinguished naturalist, "can be more advantageously, and at the same time more innocently, employed than that which is devoted to the study of Natural History." This is particularly true in regard to children, who have an inborn tendency to find pleasure in observing the habits of animals. Airiest and sprightliest of creatures himself, a boy will watch with intense delight the gambols of a kitten, or the graceful movements of a flock of pigeons wheeling about in the air. Oblivious of hunger and thirst, he will follow a squirrel from tree to tree, and at last return panting to relate what capital fun he had. It is much to be deplored that this bent of the juvenile mind is not properly directed, and that greater attention is not bestowed upon the development of the faculties of accurate observation. There are various ways in which this can be accomplished. In other civilized countries, parents take their children to zoological gardens and museums, where the young ones make early acquaintance with various kinds of animals from different parts of the world. They are imperceptibly led to compare living forms and study their characters. Nutting and shell-gathering are foreign to us, and we have no idea of the mirth, enjoyment and lessons of these healthful excursions. There, children are encouraged and helped to make their own collections of Natural History objects, and to form miniature museums, aviaries, and aquaria for the study of animal life, and other phenomena of interest. Abundant facilities for similar pursuits exist in this country also The two things needed are will and interest. We have our museums and zoological and botanic gardens, and every school-boy of every village can have his own *aquarium.*

This high sounding name need not terrify you, gentle reader. It has nothing to do with the " conventional and costly structure suggestive of plate glass, and elaborate metal and rockwork." It is quite a simple thing—a tub or a *gamla* if you like, with a few gallons of water, some tank-weeds and sand; tenanted by freshwater fishes, mollusks, and crustaceans. Such a collection cannot but be an object of absorbing interest to youthful minds, affording them early opportunities for observation and contemplation. Besides the instruction and amusement which even such a primitive aquarium is likely to afford, it exercises a certain amount of disciplinary influence upon the minds of young men in that it accustoms them to learn to take trouble about little matters, and to bestow attention on details.

To establish an aquarium, the first essential requisite is to procure a strong large *gamla,* or a tub. The former should always be preferred. A small quantity of coarse sand, or, better still, some gravel, a few stones, or, in their absence, one or two partially vitrified bricks (Jhama) should be collected and placed in the *gamla.* It is always better to expose the sand, gravel, or stones to boiling water for a few minutes, and to scrub and wash them thoroughly before filling in the *gamla* with water. Unless this is done, the various decaying animal and vegetable matters adhering to them will soon rot, and render the water unwholesome to its inhabitants. With regard to water, an experienced popular naturalist recommends that the best is that which is drawn from a river, and next to that is the water of a pond. Although this recommendation is meant for another country with different climatic conditions, it can be followed in India also.

Stocking of an Aquarium. The next important point, after having filled the *gamla* or aquarium with water drawn either from a river or a pond, is to stock it. To begin with, select some suitable aquatic plants. They will add much to the beauty of an aquarium, and afford shelter, amusement, and food to the fishes and crustaceans, and instruction to the owner. Some common plants are here indicated, and one or all of them may be used according as the place is large or small.

Sheola (*Oscillaria amphibia*), must be familiar to every Indian student who has ever bathed in a tank with old brick-steps, which are generally covered by a thick slippery layer of this species of Alga. They are filamentous or thread-like in structure and of microscopic smallness. The extremities of the filaments have a peculiar worm-like motion, which is best seen under water when it is slightly disturbed.

Pata or Halla (*Vallisnerea octandra* Willd) is another aquatic plant which may be advantageously employed for our purpose. It is grass-like in appearance, and grows abundantly in tanks and freshwater pools all over Bengal. A small tuft should be potted in a flower pot for immersion in the water.

Tokāpānā (*Pistia stratiotes*—Willd) is far too well-known a weed to require any description. A few of these weeds floating on the surface of the water will add to the beauty and utility of an aquarium.

The common Shusnisāk (*Marsilea qudrifolia*—Roxb.) is another interesting plant which will amply repay the trouble of cultivating it in a small pot for the object indicated above.

Live stock for the Aquarium. As already mentioned, materials for stocking a small aquarium, or to use a less pretentious word, a *gamla* or tub, are abundant everywhere in India. Our green lanes, ponds, tanks, swamps, and rivers teem with animal life from which selections may be made. Some trouble in searching for and procuring them is, of course, unavoidable; but it will be amply repaid by the pleasure and instruction which the objects will afford.

Pond snails and fresh-water mussels. Within twenty yards or less of one's homestead, there may be a pond or tank full of pond snails and freshwater mussels (*Gügli, Sāmuk, Jhinuk*). Fish up some and place them in the *gamla* or tub among the vegetation. They are objects of great interest, as will be found when observed at leisure.

Fresh-water sponges. Large trees such as banyan,

aswathwa (*Ficus religiosa*—Willd), or, for that matter, piyal (*Buchanania latifolia*—Roxb.) with lower branches overhanging swampy hollows, are not uncommon in villages or outskirts of towns. During the rains such swamps are invariably flooded, the water rising as high as, or higher than, the drooping branches. Floating substances consisting of straw, leaves, and rags are caught up by them, and remain suspended long after the flood subsides, until wind and rain scatter them abroad. Among these familiar *debris* are occasionally found other objects, which, owing to their superficial resemblance to the former, often elude observation. They are popularly believed to be shrimps' nests, owing probably to the fact that these small crustaceans sometimes nestle among them. Yet, among the peasantry of India, there are men here and there who are more observant than their neighbours. They will tell us at once that those irregular shapeless masses are some uncanny objects which it is not safe to touch. On examination they will be found to be *spongella* or *fresh-water sponges*, composed of some soft substance supported by a frame-work of very minute needle-like tubes, which, when touched, quickly penetrate the pores of the skin and cause pain.

The fresh-water sponges swarm with minute eggs or "gemmules," as they are called in zoological language. These are the seeds or germs of future sponges and are enclosed within a hard outer shell. At the beginning of the rains they are set free and fall to the bottom of the water. By the time the floods begin, they burst their covering and escape to find supports in drooping branches, floating sticks, or even in tufts of grass, and to take root upon them and multiply. Dirty old ponds and swampy paddy fields are the congenial breeding-grounds of these simple animals. A small mass of them will be an interesting object for the aquarium.

While looking for the *spongella* in swampy paddy fields, we come upon a small shoal of fish floundering in the liquid ooze. They are all familiar to us, and some even known by name, such as the *Khalisa*, the *punti*, the *tengra*, the *magur* &c. Any or all of them will do for an aquarium, except that, if the magur is adult, it should not be kept with other fish.

V.
EXTRACTS FROM AN ANONYMOUS JOURNAL.

Students' boarding houses have become a regular institution in Calcutta. Within a certain radius of every school and college in the central division of the city, there are sure to be one or two such clubs. Some among them have earned a sort of distinction as compared with others; and, all things considered, such distinction is not undeserved. There are boarding-houses, for instance, which can claim as their inmates for a succession of years a number of the most distinguished members of that worthy race—the students. No wonder then that such clubs, distinguished alike intellectually and morally, are never dissolved, and their traditions never die.

No 54, Carpenter's lane, was a club of some such repute. A glorious band of undergraduates having just vacated it for a better and more commodious tenement, another batch of young men took occupation of the house in the latter end of January 1869. It was an old-fashioned structure, erected like most other buildings in Bengal without the least regard to any style of architecture, ancient or modern, but with particular reference to the demands of a growing joint-family. Be that as it may, it was admirably adapted for a mess for students. That the outgoing boarders should leave a legacy of dust and dirt for the incoming ones to sweep and clean was quite in accordance with the eternal fitness of things. During the process of tidying up which therefore necessarily followed, a locker was opened, and its contents, consisting of heaps of old newspapers, torn leaves of worm-eaten books, and other miscellaneous *debris*, were consigned to the nearest dust bin. It so happened that one of the members, who had gone out on business in connexion with the new establishment, observed, while returning home, that a number of street boys were busy annexing the papers in the bin, and one of them was tearing the leaves off what looked like a foolscap-size Letts's Diary. There was something so inviting in that book with its dirty cover and neatly written pages that it naturally roused the curiosity of the student to inquire into its contents. He, thereupon, gently tapped the boy on the shoulder, and asked him to be allowed to have a look at the book. Not understanding the meaning of the intrusion, the boy stared vacantly for a few seconds, chucked the book at him, and ran away as fast and as far as his lithe little legs would carry him. Minutes, quarters, half-hours passed; yet, oblivious of time, hunger, and thirst, our student stood by the side of the bin, reading that mysterious manuscript. A dust-stained label on the cover of the journal bore the following title, "*Notes and Memoranda on miscellaneous subjects.*" None of the pages were signed or even initialled, so that it was absolutely impossible to discover the authorship of the manuscript. All conscientious attempts in this direction having failed, it was unanimously decided by the members of the Carpenter's lane Club that the journal should remain in custody of G. C. B. to whom the credit of rescuing it from an ignoble ending belonged, until such time as the advisability or otherwise of its publication should be resolved upon. That time having now arrived, the self-elected trustees of the anonymous journal have entrusted us with the publication of it, or of such extracts from it as may be deemed expedient. As for the "*Notes and Memoranda*" themselves, they are evidently the production of a young student, and are therefore interesting and agreeable, their literary imperfections notwithstanding. They introduce us into the thoughts and pursuits of a boy, and help us to follow the bent and development of his mind from boyhood to youth.

1. EARLY LESSONS IN ORNITHOLOGY.

As a boy I was very fond of birds, and, truth to say, spent more time in their pursuit than in the village school. Numerous were my escapades, but the pleasure of observing a real live bird and the hope of possessing a new one ever impelled me on to new exertions. Various contrivances were devised and adopted by me in trapping the commoner kinds. I found crows most wary and difficult to capture, and rock pigeons stupid and, therefore, easy to secure. Rock pigeons have become semi-domesticated in most parts of India; and as their presence about a homestead is considered lucky, they are generally allowed unrestricted freedom in villages. This accounts for their unsuspecting nature. Often have I caught them, and having tied some tiny bells to their feet, let them go. On one occasion, now many years ago, I saw a solitary pigeon picking up some corn or rice on the floor of a small untenanted room in our house. An irresistible impulse to possess it took hold of me, and I shut the door and flushed the bird, which in mortal fright dashed about from one corner to another. I did not notice that there was an old window with a gap wide enough to allow the bird to escape. It was however, too late; the pigeon had already gone. But as ill luck would have it, when struggling to get out, it was pounced upon by a large tom cat, one of those half-wild, half-domesticated creatures which prowl about every house in every village in Bengal. The poor bird lay bleeding and lacerated; its convulsive flutterings every now and then indicating great agony. The sight almost paralysed me, but there was no time for reflection. My greatest anxiety then was how best to conceal the fact of the accident from my mother, who, tenderly as she loved me, would have been shocked to hear that I had brought it about. Having therefore gently picked up the bird I ran towards the farthest end of the garden, and laid it by the side of a small well. As I was preparing to wash the wound, I was overjoyed to see dear old Janardan before me. He had already taken the bird in his hand and was examining its injuries. What a relief it was to me to have the company and help of Janardan at such a moment of need and perturbation! Without, however, wasting a moment's time to ascertain the cause of the accident, he bade me fetch some turmeric paste from the kitchen, a commission which I executed with the greatest alacrity and quickness. Having carefully washed the wounds, Janardan applied some of this paste over. It was arranged between ourselves that the bird should remain with him in his hut. I was extremely delighted to find the next morning that the patient was very much better, it took food from Janardan's hand, but refused it when offered by me. Janardan said it was yet angry with me, and I verily believed he was right. Gradually its avian heart relented towards its unintentional tormentor, and it became as familiar and confiding to me as it had already been to Janardan. In the meantime he had related with much good sense, the story of the accident and of my concern and solicitude for the recovery of the bird, to my mother, who gravely admonished me not to behave in that manner again, and to be always kind and considerate towards animals. The pigeon having now completely recovered, it was installed as a member of our family, and in a short time learnt many tricks and performances under the able and skillful tuition of Janardan. It constantly afforded much amusement to the visitors to the house; and I remember it being declared by competent observers that it could distinguish acquaintances from strangers, when the company were small. It had learnt that my mother was the mistress of the house, and that all appeals for food and drink should be made to her; so that, when hungry it always came wheeling and dashing to her side and began cooing in an absurdly pitiable way. As the bird never objected to our manupulating it as we liked, Janardan took advantage of its trustful nature to give me some useful lessons on elementary *ornithology*. He would ask me to count the feathers of its wings and tail, and to observe that the large wing and tail feathers were overlapped by sets of smaller feathers above and below. What a dunce I was not to be able to answer that wings were used for flight. To demonstrate this fact Janardan playfully tied up the quill feathers with a piece of string and let it go. Oh no, it could not fly! When I asked him about the function of the tail feathers, he in his usual pantomimic way told me to find it out for myself by observing the flight of birds.

The crop of the pigeon puzzled me much. I thought it was its stomach, as I could feel the grains it ate, but my juvenile wisdom could not solve the mystery of its being situated at the breast. Janardan however set me right. He explained to me that it was a bag formed by the dilatation of the gullet or tube through which food passed into the stomach, and that it served as a reservoir for food hastily taken. I have since learnt that it not only serves as a reservoir but by its secretion moistens and softens the food which consists mostly of grains of various kinds.

2. "IN MEMORIAM." JULY 18TH 18—

To be able to read well with a clear voice and correct accent is an essential part of a good education. But however clever an Indian student may be otherwise, he is hopelessly deficient in this respect. To remedy this defect as far as it lies in his power, our good Professor ———, has introduced the admirable system of making each of us read aloud by turn, selections from standard works in prose and poetry. I take shame to myself that the system proved very disastrous to me yesterday. It was my turn to read the introductory portion of Tennyson's "In Memoriam." All went well for a short time until I came to read

> "Forgive my grief for one removed,
> Thy creature, whom I found so fair,
> I trust he lives in thee, and there
> I find him worthier to be loved."

> "Forgive these wild and wandering cries,
> Confusions of a wasted youth;
> Forgive them where they fail in truth,
> And in thy wisdom make me wise."

My voice became choked, and eyes moistened; all my latent grief for a dear friend lately lost, welled up. I became confused, and despite my efforts not to make a fool of myself, broke down completely. That friend was Janardan, a kind-hearted wiry old man, full of helpful ways for young and old of every village for ten miles round. Like mine to-day, many a yearning and grateful heart must be sighing for him these last two months. He was, however, the especial delight of children who found in him a constant yet wise abettor of all sorts of innocent juvenile sports and pastimes. He it was who whittled my first *danda guli*, and initiated me into the mysteries of that healthful village game. The first bird that I could call my own was his gift. It was a fledgeling *Gang salik* (Bank myna) just taken out of a hole in the steep bank of the Bhagirati.

IMAGES OF THE PAST. In calling up images of the past, certain river-side scenes which took hold of my childish imagination come before my mind's eye in all their freshness, and with all the details clearly defined, as if, I were actually gazing at them now. I find the broad and impetuous river dwindled into an insignificant stream. An abrupt and precipitous bank riddled with countless holes and supporting a rugged bluff flanks it on one side; while on the other a vast expanse of sand spits slope gently down to the water. All is quiet, except a few straggling *gangsaliks* hanging about the holes, or a solitary tern skimming the surface of water on the opposite shore in quest of mullets, which affect shallow water. Beyond the low lying flats of white sand, strewn every-where with broken shells of fresh-water mollusks, the spits form a somewhat higher bank supporting a luxuriant vegetation of tamarisk *(Jhao)*. Here and there are depressions, like miniature lakes, some small and some large, but all abundantly fringed with tall reed grass. A couple of Cormorants are disporting themselves in one of them, while Snake-birds are playing hide and seek in another. Further inland, and dimly visible against the horizon is a long and unbroken line of trees and shrubs overhanging what look like low rugged banks of some river. No doubt they are the indisputable evidences of the changeableness of the river, of which the channel flowed past them not so very long ago.

The aspect of nature becomes quieter, and the shadows begin to lengthen: there is no sound, not even that of a leaf falling or grass rustling. Suddenly a flock of birds flies past overhead us with a loud but transitory whirr; another follows, and yet another. They congregate at the bluffs, and we recognise them to be our friends the *gang-saliks*. The peaceful solitude of the river-side is now transformed into a scene of much excitement and unrest, caused by the chirping, jostling, and fighting of vast flocks of birds, their number augmenting every minute by the arrival of fresh contingents. There is a lull in the confusion and we hear the weak, subdued screaming of innumerable chickens. I wonder how each bird finds it own hole! Merciful heavens!! What is that owl sitting there for, on that broken bough of the *bael* tree? An indescribable sensation of chill creeps over my body at the sight of its magnificent pair of furious eyes.

That night my sleep was much disturbed. I dreamed of the owl, and of its awe-inspiring eyes; of its bloody work among the birds, and of their shrieks of agony.

3. THOMAS PERIWINKLE.

As I was sitting in my room yesterday afternoon deeply absorbed in reading A's account of his journey from Calcutta to Aden *en route* to England, I heard some energetic bangs at the front door which almost shook the building. Before I could get up to ascertain the cause, a shriek accompanied by loud slamming of the door had brought half the inmates of the house to the spot. On looking out from the window which overlooked the lane I was relieved and overjoyed to see Tom Periwinkle, standing near the door convulsed with laughter. Our good old maid servant, Bhaba's mother, in spite of her fifteen years' residence in Calcutta, be it said to her credit and honor, and to our unspeakable good luck, is yet as simple and unsophisticated as she was on the day when she first left her village to seek service in this city. The dread of white men which possessed the minds of village maidens in Bengal in the early days of the Honorable John Company's rule is still extant. No wonder that a simple creature like our maid servant should be scared at the sight of a *gora*—a whiteman at such close quarters. Besides, the nondescript appearance of Tom with a parrot on his shoulder, a cageful of small birds dangling from his right hand, a large packet rolled up under his left arm, and a long German pipe with a grotesque-shaped bowl in his mouth, was enough to frighten anybody. The excitement of the occasion having subsided, and the first greetings over,

Tom was conducted to the room of S. C. M. for whom the birds and packet were intended as presents. Good humour is catching, and young men, especially young students, are particularly susceptible to its influence; so that, within five minutes of Tom Periwinkle's arrival the place was ringing with laughter, and over-flowing good humour of half a dozen choice spirits.

OUR FIRST ACQUAINTANCE WITH TOM PERIWINKLE. It was late one cold, foggy December night. The streets were almost deserted. Occasionally a belated pedestrian wended his way homewards; or more rarely, a sleepy policeman muffled up in his great coat could be seen leaning against the lamp post, and appearing through the mist of night, more like a shadow than anything else. S. C. M. was returning from a marriage feast at a relation's house in the northern part of the city. While at the crossing on the N. W. Corner of Cornwallis Square he thought he heard some one sobbing. He paused, but the only sound that fell upon his ears was that of a ricketty hackney coach that jangled laboriously by. He was about to pass on, when he heard the sobbing again. Indistinctly visible through the dull vapour of the night was the shadowy outline of a form sitting on the foot-path. True enough! There sat, sad and forlorn with his back towards the railing of the square, his legs drawn up and his arms crossed over the knees, a European lad, apparently not more than eleven or twelve years old. His head was bent forward and resting on the crossed arms. He was evidently in great anguish of mind. His soiled boots, dishevelled locks, and meagre garments, were naturally suggestive of a tramp. But all such ideas vanished from my friend's mind when in response to his sympathetic enquiries, the lad held up his head preliminary to speaking, revealing a pair of beautiful hazel eyes much swollen from weeping, and a comely oval face—sure index of a guileless, honest heart. So overpowered was his boyish nature at the sympathy shown by my friend towards him, and at the prospect of succour at hand, that all his pent up feelings burst forth at once, and it was sometime before he could stammer out a few words in reply to my friend's enquiry as to who he was, and what he was there for.

But all that S. C. M. could gather from the broken sentences intermittently jerked out—he was still sobbing —was that he belonged to some steamer in port, and that he would "catch it" from the Doctor. As the night was far advanced, and the boy was badly in need of food and rest, my friend proposed to take him to his lodging and to give him both. It required very little persuasion to induce Thomas Periwinkle, to accept the invitation and accompany him to his club. Having arrived there and found an unoccupied bed, Tom straightway went to lay himself down, and in five minutes was sound asleep, so that when refreshments were brought, it was with utmost difficulty that he could be made to take a cup of tea and a few biscuits.

Tom Periwinkle's Adventures in the Streets of Calcutta. One afternoon, close upon two years ago, Tom Periwinkle stepped ashore with a light heart. He had some elderly companions with him. They were soon inveigled into gin shops by some of those low trickish fellows who loiter about the quays, and the precincts of the Sailor's Home, and who on pretence of befriending Jack invariably manage to transfer the contents of his pockets into their own. Tom had honest scruples about entering such places, and therefore parted company with his friends and wandered about at his own sweet will. Guileless and honest, full of rollicking fun, Tom was a mere lad, and therefore, absolutely incapable of protecting himself from the designing villany of such scoundrels, who consider jack tar their legitimate quarry. He soon allowed himself to fall into their clutches. What the end was we have already seen. Cheated of his money and robbed of his coat and hat, Tom, hungry and thirsty, had trudged the lanes and streets of nearly half the city in the vain attempt to find his way to the shipping, until at last he emerged into the square where he sat forlorn and desolate that foggy December night.

Thomas Periwinkle's Characteristics. Tom Periwinkle took to us very kindly after the adventure of that night, and has ever since been a great friend of ours. Although a mere boy, not more than thirteen or fourteen years old, he is already a citizen of the world, can talk several European languages, and besides possesses a smattering of Hindustani which he has picked up from the Lascars. He is as gay and contented while trimming sails, sitting astride the yard arm as he could be if playing in the streets of London, or beside his mother's door. His manner is very unaffected and his demeanour full of heartiness; it is therefore quite natural that he is a universal favorite with the officers and the crew. Cheerful and vivacious, he is ready to lend his hand to any work at any time. Free and joyous like a bird and full of fun, he runs up the rigging, or lets himself down into the hold. He has already seen many countries, and, boy as he is, delights in telling stories about the places he has visited. We have nicknamed him "Geography," as a compliment to his knowledge of the subject gained from practical experience. We have found it ten times more interesting and profitable to listen to his simple boyish tales, descriptive of places and people, of animals and plants than to read the dry details of a geographical reader. Good Dr. Mere-weather, the ship's Surgeon, is Tom's guardian. He is a keen naturalist, and has succeeded in infusing the same taste into his ward, who has already proved himself a capable hand in skinning, preserving and mounting natural history objects.

* * * * * * *

I was not aware until to-day that the birds—a Grey Parrot, and some Finches, were for me, and the packet containing Wordsworth's Proseworks were for S. C. M. They are from Tom's mother and we are grateful to her.

Any trifling service that we were able to render to her darling son has been a great deal more than repaid, by the pure friendship of Tom, which we have since enjoyed, and by her own letters, conveying warmest sentiments of affection and good will towards us. Apart, however, from their intrinsic worth, which is much, the gifts are valuable to us for another reason. They are unmistakable proofs of the good understanding which has gradually developed between individuals of two races who are unfortunately always prone to misunderstand one another. I value the gift much, as it is accompanied by an elaborate note on the natural history of Grey Parrots, which Tom has taken the trouble of drawing up with the help of Dr. Mereweather for my information.

THE GREY PARROT.*

It is a beautiful bird. The general colour of the plumage is pearl grey; the feathers of the head, neck, and belly have lighter margins. The colour of the tail feathers is bright red; the beak black or dark grey; the legs ashy. The male and female are almost alike. The Grey Parrot learns to talk well; the male, it is said is usually the more fluent talker than the female, but the few words the latter learns are pronounced with great distinctness. It is evident from the various anecdotes current about this bird that it is endowed with a large amount of intelligence.

---

* Complied by Thomas Perwinkle, with the help of Dr. Mereweather, from Gzaen's Parrots in Captivity.

ANECDOTES OF GREY PARROTS.

A Grey Parrot in possession of a medical man had learnt to say, "open the door, and call the doctor," but occasionally it reversed this order, and shouted out, "open the doctor, and call the door." This clearly shews that although it attached a certain meaning to the sentence it used, the several words of which the sentence was composed conveyed no idea to its mind.

Once upon a time a lady of high rank had a bird of this species. It was given to her by a man who had for a long time lived in the East Indies. It could, therefore, talk only Dutch. In a short time, however, it learnt both German and French, and these three languages it spoke very distinctly. It had wonderful memory, besides being very attentive, so that it often picked up expressions which *had never been used before it*. When thirsty it would say "Polly wants kluk kluk," meaning drink; if hungry "Polly wants something to eat." If food or drink was not forthcoming, it exclaimed: "But Polly must and will have something to eat!"

The bird was not fond of strangers. Those who came to hear it talk were generally disappointed, or had to gratify their curiosity by hiding themselves before their curiosity was gratified. It was, however, easy to win over its confidence and affection by kindness. The bird had a certain sense of humour. Once a military man—a Major whom the bird knew well, visited it and introducing his walking stick inside the cage said: "Jump on the stick Polly, on the stick!" Polly got angry, and suddenly burst forth laughing saying: "Major, jump on the stick, Major!" A son of the family to which the parrot belonged, was expected home after a prolonged absence abroad. The event was naturally talked about among the household. George, for such was the name of the son, arrived late one evening, when Polly was already sleeping in its cage. After the excitements of the first greetings were over, George sought the general favourite and lifted a corner of the cover of the cage: "Ah, George, art thou there? that is nice, very nice," were the words of welcome and recognition uttered by Polly.

* * * * * * *

The Grey Parrot is a favourite cage bird, but is no where so common as in London. Large numbers are continually imported to Europe from their native wilds.

The home of the Grey Parrot is the western coast of Africa, extending for some distance into the interior. It is common in the Gold Coast and adjacent islands. There is some peculiarity in the distribution of these birds in these parts; for instance, it is very abundant in Princes' Island, but not one is to be found in the neighbouring island of St. Thomas. The explanation of this lies in the fact that, the large number of kites which inhabit St. Thomas's Island prevents the Grey Parrots' coming and settling there.

A distinguished observer gives the following account

of Princes Island—the Paradise of the Grey Parrots, and of the habits of the latter in that place.

"On Princes' Island there is a very lofty mountain, reaching some 1200 feet above the level of the sea, and called by the natives 'Pico de Papagaio,' or Peak of Parrot. On the slope of this mountain, and extending far up its sides, is a magnificent forest. The trees are of great size and height, and their trunks and branches give support to the *lianos* and other climbing plants, which hang about them in rich luxuriant folds. The density of the forest is so great that it is only with the utmost difficulty and toil that the explorer can force a passage through it, while to the Parrots, who come there every night, it presents no obstacle, but gives them, under the shelter of its thick foliage, a secure and pleasant resting place."

"As sun-set draws on, the Parrots may be seen in parties winging their way for the mountain from all sides of the island, and on reaching it, take their places on the trees. Approaching troops acquaint their fellows of their coming by a loud whistling. Those who have found an approved resting-place warble and whistle as long as day light continues, but as darkness closes in the noise gradually subsides, and all becomes hushed. Occasionally, however, a few sounds may be heard at intervals after dark, which most probably proceed from some belated bird seeking a place or a quarrel : sometimes in the dead of the night the whole colony is thrown into an uproar, occasioned, I believe, by the visit of cats or of some predacious animal."

* * * * * * *

The *Food of Grey Parrots.* In their own country and in the wild state, the Grey Parrots feed on fruits and grain; in captivity they thrive well on the same kind of food varied by such delicacies as bread and milk, biscuits, boiled potatoes. They are particularly fond of sweet things, such as dates or lumps of sugar &c. Tenderness and affection for young ones, not their own, are not uncommonly manifested by them. When large numbers of young and old Grey Parrots are caged together, the older birds frequently feed the younger ones, which without this attention would probably die.

4. OF PARROTS IN GENERAL AND OF INDIAN PARROTS IN PARTICULAR.

Since reading Thomas Periwinkle's interesting account of the Grey Parrot, I have paid some attention to this family of birds. They all seem to resemble one another in having four toes on each foot, two of which are turned backwards. This arrangement gives them great power in grasping or climbing. The feet answer every purpose served by hands, so that these birds can grasp or hold fruit or nut in one foot whilst holding on to a branch with the other. But the formation of the foot is not favourable to walking, which, they therefore avoid as much as possible, living almost entirely on trees where they find their sustenance. The bill is short, thick, and strong; the upper half or the "upper mandible," as it should be called, is much curved, and has an acute hooked tip: it is movable, and overhangs the lower mandible. The latter is short and obtuse, rather square. Their powerful bill helps them much in climbing. The tongue is fleshy and rounded, and in some, provided with brush-like bristles. The flight of the parrot is moderately fast.

Parrots, as a rule, are very cleanly in their habits, and if not actually engaged in feeding, or other works of social or individual economy, they employ their time in toilet operations, which generally consist in carefully preening all the feathers. On one or two occasions I have witnessed a flock of Ring-necked Parrakeets enjoying a bath in a small ditch of clear water by the side of a field of ripening wheat corn. In captivity, also, they are fond of bathing, and it is therefore advisable to provide a caged parrot with a shallow pan of water during the day. They eat much, but waste more, apparently in wanton mischief. Any one who has ever witnessed a parrot feeding in captivity must have noticed, that it takes a beakful of grain, paddy or hempseed for instance, and lodges it under the tongue, takes it out again grain by grain, separates the husk from the grain, and eats until its hunger is satisfied. It does all this with marvellous quickness and method.

I lately saw a remarkably large black Cockatoo, which belongs to the same family as parrots, with a fine crest and a large beak, in one of the bird shops at Territibazar in Calcutta. It fed on large hard American nuts, which the owner said he could with difficulty break with a hammer. Yet I saw the bird cracking them with ease and extracting the kernel. Parrots are evidently fond of the seeds of sunflower, as I have often seen small flocks of Ring-necked Parrakeets, our common *Tiya*, frequenting gardens during the winter to feed upon the ripening seeds of sunflowers, and again during the rains, when the *Radha Padma*, an allied species, flowers and sets seeds. At breeding times they generally cut a circular hole in the trunk of some tree, and lay several white eggs. I have, however, seen Ring-necked Parrakeets breeding in small circular air passages in walls of old buildings. Although there are several species of parrots found in India, none of them is remarkable for brilliancy of plumage. As far as I have been able to find out, parrots of gorgeous plumage appear to inhabit mostly the Malayan Archipelago, extending as far as Australia. America too can boast of some bright-coloured species. In the bird market of the Territibazar I have seen parrots from almost all parts of the world, America, Australia, Africa, the Malayan Archipelago, India, Burmah, Ceylon, the Andaman and Nicobar islands. Europe does not appear to possess them. Ordinarily, parrots are fond of living together in large flocks, which often rend the air with their loud screams.

Of the various species of Indian parrots, the following are generally made pets of :—

The *Chandana** is found in the forests of Northern and Central India, including Nepal, Bhutan, the Punjab &c. It is said that a parrot of this species was taken to Europe by Alexander the Great and hence the bird is sometimes called the "Alexandrine Parrakeet."

The Tiya *(Tota Pakshi)* or the Ring-necked Parrakeet,† is one of the most common and familiar birds in all parts of India, from the foot of the Himalayas to the extreme south extending to Ceylon. They are more commonly found near towns and villages, than in thick forests. While we were staying at Monghyr during the last winter, we saw vast flocks of this parrot roosting together every night with other grain-eating birds in a small mango tope; and the noise they made at sunset and sunrise was indeed deafening. I am not sure whether other birds did it or not, but on several occasions I saw the parrots bringing, at day time, ears of Indian corn to their roosting place, probably to devour at leisure, and unmolested by others. It is customary with every grocer of every rank in India, especially in Bengal, to have a parrot as a pet to enliven his solitude; and whether from choice or accident, in nine cases out of ten this pet happens to be a Ring-necked Parrakeet. He spends much time and trouble upon its education, teaching it to repeat the names of *Radha* and *Krishna* or *Kali* and *Doorga* according as he is a Vaishnava or a Shākta. In spite of his well-meaning attention to

---

* Palæornis nepalensis, *Hodg.*
† Palaeornis torquatus, *Vigors.*

his pet, I must confess I do not like his keeping it in a small circular iron cage, where its tail feathers get broken; or worse still, chained to a ricketty swing. Besides being very small, the iron cage must get very hot during the fierce mid-day heat of summer, and again very cold in raw wintry nights.

The *Fariadi*,* or complainer, so called probably from its plaintive note, is a beautiful parrot, with its whole head and face of a delicate roseate colour, and tinged with plum bloom at the sides. It is found almost all over India, extending into the Himalayas, Assam, Burmah, and Ceylon. As a pet it is less prized than either of the two preceding species, or *the Kajla*, mentioned below, although large numbers of them are sold in Bengal and other Provinces. *The Madna Tota*,† or *Páhari Tiya* comes from the lower ranges of the Himalayas, and the hill ranges of Assam and Sylhet.

The Madan-gour Tota inhabits the jungles of Malabar Coast, and up to about 5000 feet, or upwards, the slopes of the *Nilgiri* (Neilgherries). Both the above species are handsome birds, and are frequently kept as pets, but their intelligence does not seem to be on a level with that of the *Chandana*, the *Tiya*, or the next species.

The *Kajla*‡ :—This parrot is a great favourite with most bird fanciers in India, and it thoroughly deserves the attentions bestowed upon it. It has a much sweeter

---

* Palaeornis rosa, *Jerdon*.
† Palaeornis schisticeps, *Hodg*.
‡ Palaeornis javanicus, *Osbeck*.

and softer call note than that of the *Chandana* or the *Tiya*, and is decidedly more intelligent and affectionate. It is found in the sub-Himalayan regions, and is also abundant in the jungles of Assam, Sylhet, Hill-Tipperah, extending to the Malayan Peninsula and Java. It is exported thence for sale to the other parts of the country.

THE STORY OF THREE KAJLA PARROTS.

The only son of a poor woman, and she a widow, lay sick unto death. The mother had done all she could towards alleviating his suffering and saving his life, and was ready to lay down her own to preserve that of her son. But slender was the hope of recovery! Doctors and Kavirajs had given the case up. It was by the merest accident that Svamiji, Janardan's great master and spiritual guide had come to see the patient. Though not a physician by profession, Svamiji had a thorough knowledge of the Indian medicinal plants, and of their habits and haunts, as well as a fair insight into the Hindu system of medicine as practised by the ancient masters of the healing art in India. As it were by an unerring instinct, he would sometimes come to the determination of a disease, and the selection of proper remedies; at least such proved to be the case on the present occasion. Having given some simple directions as to diet and hygiene, and having pronounced his benedictions upon the patient, he left his bedside and took the road, equipped as he was, in the simple garb of a mendicant, to the nearest Railway station, distant twenty one miles. On the seventh day after their departure Svamiji and Janardan his faithful disciple, were wandering about some forest in Hill Tipperah in search of a rare medicinal herb. During one of their afternoon ramblings through a forest glade Janardan's attention was arrested by a soft plaintive screaming sound, which came mingled with the clear piping note of a Kalij pheasant. It did not take Janardan five minutes to trace the screaming to some young and unfledged parrots which had their nest in a *gahmar* tree. One of the young birds lay dead at the foot of the tree, the rest crying and crowding at the entrance of a circular hole where the trunk forked some eight feet above ground. It was evident from a few blood-stained feathers that lay scattered about the place that the parent birds were dead, either killed by some cruel sportsman, or preyed upon by some rapacious birds. Janardan did what any other kind hearted man would have done in the circumstances. He rescued the famishing young creatures from their perilous position, and busied himself for the next half hour in feeding them, and making them comfortable. But his best endeavours proved unavailing, as the poor things continued crying and gaping without being able to take any nourishment. There was something wrong somewhere. Janardan was however, a resourceful man, not easily discouraged by difficulties real or imaginary. Having taken a somewhat long wink with both eyes closed, he took out a small handful of parched gram from his inseparable knapsack or *Jhola*, and surreptitiously put it into his mouth, and began chewing the same—until it was reduced to a fine pulp. This ingenious device of Janardan's succeeded admirably, as the birds began eating out of his mouth just as they used to do from that of their parents. The young parrots holding on to Janardan's abundant and flowing beard, and eating out of his mouth was a ludicrous sight indeed, and moved even the grave and sedate Svamiji to laughter.

On his return Janardan at once disposed of two of the birds, presenting one to the patient for whose benefit the journey was undertaken, and the other to the first juvenile friend he met. I am told that he was ready to give away the third to statisfy the importunity of another boy friend of his, but the young bird had taken so kindly to him that it positively refused to part company with him.

Janardan's idea of bird-fancying differed considerably from the orthodox system in that his numerous pets, comprising robins, bulbuls, orioles, sunbirds, cuckoos, wagtails, crows, and magpies had never known what captivity was, since the day when their education was supposed to be completed, and yet they were ever ready to obey his beck and call. When after an absence of a fortnight or more, Janardan returned to his Arcadian retreat, the *doyal* birds warbled their rich sweet songs of welcome, to which the bulbuls responded in chorus. From behind the leafy bowers came the soft and sweet notes of the shy and timid oriole; but the tiny and graceful sunbirds were very demonstrative: some of them hovered round him in the air, others flitted about from bush to bush in great glee, chirping incessantly all the while. The koels and the crows were out foraging in another part of the wood, and were not apprised of Janardan's arrival, until the solitary magpie had made the grove echo and re-echo with its loud and clear metallic whistling note, so different from the ordinary harsh cry of the species. Janardan was a perfect adept in imitating the songs and call notes of birds, and talked to his pets in their own language, and signified his pleasure at seeing them happy and joyous. He was very happy too, if one who knew no unhappiness could be called happy, and stood contemplating the birds till night set in, and spread its dark mantle over the earth, the wood, and the sky, and hushed the birds into silence. But still Janardan moved not. His thoughts had wandered from the birds to their great Creator and he was lost in praise and thanksgiving for all His dispensations.

I once asked Janardan why he never caged his pets like other bird-fanciers? His answer was characteristic. "Know you not," he said, "that the birds are formed for a life of freedom." Caged or not, it would have gladdened the heart of any man to see the confidence with which his birds approached and moved round him. The secret of this lay in his sweet lovable nature and his great sympathy which embraced all God's creatures.

To resume the story of the parrots. As we have

already seen, one of them still clung to Janardan. Although according to his notion its education was complete, he did not venture to let it have freedom at once, as the surroundings were quite new to the creature. Some sort of accommodation was therefore necessary, and a small old rattan basket being at hand, it was placed on two pegs in the mud-wall of his neat little hut, with a thick tuft of soft dry grass as a bedding. Thus housed, the young bird felt very comfortable and happy, and flew in and out as it listed, and in a few days made friends with other birds. About four months after the events narrated above, business took Janardan to Basantapore, a considerable village on the right bank of the Bhagirathi.

It was evening. The western sky was glowing with crimson and gold; the still waters of the sacred Ganges reflected the colours in tenfold beauty. The village common and everything around were bathed in a cheerful glow of exquisite loveliness. Anon, the golden rays of the setting sun quivered and danced upon the river breast, and the colours faded away. The glory of the sunset deserted the fields and meadows, but yet lingered upon tall *Hijul* trees (Barringtonia), which, with their clusters of crimson flowers formed a momentarily resplendent back ground to the prevailing dullness of the common. Birds that feed during the day were returning in flocks to their accustomed roosting places, transforming every road-side tree into a scene of great excitement. Night-herons, and for that matter, other birds that forage at night were preparing to make excursions to happy hunting grounds, and we wished them good luck. There was a spotted-owl quite near; Janardan heard its characteristic hooting, and stopped to ascertain the wherabouts of the bird, which, from its safe arboreal retreat, watched, with no pious intention, a pair of field mice that were disporting themselves in the seclusion of a bushy tuft of spear grass. Perched on a high grassy bund of a mulberry field, the *Bharat-paskhi* (Sky-lark) twittered its closing song, rising every time into the air as it sang. The greedy Brown strike was casting a longing look upon a beetle which crept up the stem of a small castor plant. It made a dart, secured the prey, and pecked at it with much vicious energy, till the poor little thing was dead, and fit to devour. With a plough and harrow upon his shoulder, driving a team of sleek bullocks before him, and enjoying every now and then a pull from his *hooka* came Faraji Sheik, a small, but substantial farmer. "*Salām Bawaji*" he accosted Janardan, but Janardan heard him not, stared this way and that, and looked confused. Though a great friend of Janardan's, and having much respect for his piety and integrity, Faraji Sheik did not understand his friend's character in all its aspects, and believed, in common with other people of the neighbourhood that Janardan was a great, but harmless, sorcerer. His notion about his friend's sorcery received a singular confirmation when a parrot came fluttering and sat upon Janardan's shoulder, and began rubbing its head against his neck in a most affectionate manner.

It has been already noticed that Sukumar, a boy friend of Janardan's, got one of the young parrots. Though good-natured and fond of birds, the lad was careless and lazy, and in spite of the remonstrances of his guardian, he would often forget to feed his pet at proper time ; as to cleaning the cage he would seldom attend to it. No pious Hindu would so much as touch food and drink, if a single animal in the house remained unfed and uncared for. From this point of view, old Devi Datta had not unfrequently to pay the penalty of his nephew's transgression by having to personally attend to the feeding of the parrot. Whatever his sentiments might have been as to the propriety of a bird fasting in his house, he did not like Polly which hated him in return. One morning Sukumar was unconsolable at finding that his pet had escaped. Whether old Devi Datt had a hand in the bird's escape or not, deponent saith not; but certain it is, that he made no attempts to recapture it. After having hovered about the place for a few days, always eluding other people's attempts at depriving it of its liberty, it winged its way towards Basantapore, where it soon made friends with a male bird of its own kind, which Sham Chand Mudi—a thriving grocer, owned as a pet. It had in the meantime developed roving propensities, and was very erratic in its movements. So that one day it would keep company with a small flock of *Goshaliks*, while on the next, it careered about with Indian rollers and magpies. On that particular day when Janardan was approaching Basantapore, it was out on a pic-nic with some Ring-necked parrots, but having seen Janardan, had followed him unobserved for about quarter of an hour, giving notice of its near presence by occasional screams which awakened memories in Janardan and puzzled him.

VI.

ROUND THE INDIAN MUSEUM.

One forenoon during the Christmas week of the year 18—a number of school-masters and Pandits from the various Presidency districts had gathered together at the Indian Museum, at the invitation of Mr. W. who had kindly undertaken to show them round its galleries. Mr. W. was then an Inspector of schools in Bengal. He was a man of wide experience and large sympathies. Teachers and students all looked up to him as to their best friend, and a friend he was in no conventional sense of the term. He was deeply interested in the spread of education in Bengal, and was always planning schemes for improving the existing systems and remedying their defects. His methods were sometimes original too. He had become painfully impressed with the fact that the lower grade teachers and Pandits—excellent men many of them—lived buried, as it were, in the oblivion and obscurity of mofussil stations and outlying villages, and were therefore practically isolated from all progressive ideas of the world; that there was no enthusiasm in them, and that they worked more like machines than like intelligent human beings conscious of the great responsibilities which their position imposed upon them. He had also noticed that being themselves deficient in the faculty of observation they failed to stimulate it among the students. With a view to remedy these and other defects, and to improve the minds and status of the teachers, Mr. W. had conceived the idea of a "School-masters' Union," and had established the system of winter and summer excursions. To defray the expenses of these excursions, a fund called "School-masters' Union Fund" was started, and to which even the poorest of Pandits very gladly and willingly contributed. The winter sessions were invariably held in Calcutta. Here the school-masters were thrown into the company of cultured European and Indian gentlemen, who one and all endeavoured to make their stay in the city as enjoyable and pleasant as it was in their power to do.

The principal feature, however, of the Calcutta session was a series of highly interesting and instructive evening lectures on elementary scientific and historical subjects with magic lantern exhibitions. The days were spent in visiting the Indian Museum, the Botanic Garden, and other places of interest and instruction. Such in brief is the history of a movement under the auspices of which that gathering of school-masters and Pandits had taken place at the Indian Museum on that forenoon of the year 18.—

BONELESS ANIMALS.

In order that they might properly understand the design or scheme underlying the animal creation, Mr. W. had first taken the party round all the zoological galleries just calling their attention to the outlines of forms of animals belonging to each class; so that when they came for the second time to the hall known as the "Invertebrate Gallery" an interesting discussion had already begun regarding the various forms of animals which they had for the first time seen in that place.

"What a variety of forms!"

"From all parts of the world."

"The vastness of the collection is perfectly bewildering!"

"Not so much so as those strange weed-like things," said Vidyabhushan, pointing towards some really very plant-like objects kept in cases against the western wall of the hall. Pandit Jadavchandra Vidyabhushan was a scholar well versed in grammar. From being an ordinary Pandit of a grant-in-aid school, he, by dint of industry, integrity and perseverance, had risen to be a Deputy Inspector of Schools, and was the right-hand man of Mr. W. The sight of those plant-like objects in a place devoted only to zoological specimens had puzzled him much.

Mr. W. who was attentively listening to the conversation and had noticed Vidyabhushan's embarrassment, explained that though weed-like in appearance they were in reality *animals*.

"Truth *is*, as they say, stranger than fiction," exclaimed Vidyabhushan.

"Let us hear something about these strange forms," cried many almost in a chorus.

"Well then," resumed Mr. W. "those weed-like objects are 'Zoophytes or Plant-animals,' so called owing to their superficial resemblance to plants. They are many-celled animals, most of which are immovably fixed, forming tree-like colonies either in the sea or in fresh water pools and *jheels*. Sponges and corals are familiar examples of these Zoophytes."

"Here is a new mystery; I wonder what that mushroom-like thing is!"

"What! a mushroom here? I thought mushrooms were vegetables."

"And so they are."

"It is a 'medusa,' just read the label there."

"So the monster is captured and exhibited."

"What do you mean."

"I mean the fabled monster of ancient times whose head was adorned with snakes instead of hair."

"Oh I understand. Those thread-like filaments hanging round its margin like a fringe resemble little snakes, and that thing is called 'Medusa' from its supposed resemblance to the head of the monster of that name. At least I remember having read so somewhere."

"You are indeed clever."

"Small service is it to know the history of the name of a particular object, and still less to see this wonderful display of animal forms, unless they are viewed in relation to one another."

"Anyhow I am curious to know what it is."

"It is a jelly-fish."

"Fish! why is it not kept in the fish gallery."

"Wait, wait. Do you see any sign of a skeleton anywhere in that medusa? I don't; but if I remember it correctly, Mr. W. told us a short while ago that fishes possess a bony frame or skeleton."

"Oh yes, I remember it. He said that while we were going round the fish gallery."

"It may be a jelly, but not a fish."

"Good, you are nearer the truth."

"Can you show me any animal in this hall possessing a skeleton," asked Mr. W., addressing the party.

"I don't see any," replied one.

"Neither do I," said another.

"Nor I," exclaimed a third.

"Do you remember having seen any animal in the other halls of this museum that has not a skeleton," again asked Mr. W.

All answered almost in a chorus that they did not.

"What does that signify?"

"What?"

(1) "That there is not a single animal in this hall possessing a skeleton, or, more correctly speaking, backbone; (2) on the other hand, there is not a single animal in any other hall without a skeleton."

After a few more questions and answers one of the young teachers suggested that this arrangement of keeping all such animals as have no skeletons in one place, and those that have skeletons in another, was probably meant to represent that *the animal kingdom is primarily divided into two great divisions.*

"Clever young man!" exclaimed Mr. W. with great delight "I am very much pleased with your answer. It shows that the faculty of observation is in you and it only requires rousing."

"All the objects then that we see in this hall," said some one among the party, "belong to one sub-division of the animal kingdom?"

"Certainly."

"How can that be, they are so unlike one another."

Hari Charan was always hasty in his remarks, and had therefore earned the nick-name of "hasty Hari Charan;" nevertheless he was an intelligent and honest fellow, and much liked by his pupils and brother teachers.

"Enlighten us please," said Mr. W. with a kindly smile.

Hari Charan was much confused with shame, and said apologetically that he was sorry to have been so hasty. (A chorus of laughter from all sides). He simply meant to say that although belonging to the same divisions of the animal kingdom, the animals there exhibited did not look as if they were related to one another; and he was therefore curious to know their relationship.

"Praiseworthy curiosity indeed!"

"All of you," said Mr. W., "must be familiar with

many animals that you always see about you. Do you see any of them here or others that have any resemblance to them."

"There is none that I can see" was the abrupt and ready answer of "hasty Hari Charan."

"Will you name some common animals, Hari Charan?"

"Goat, cow, dog, cat, &c."

"Is that all? no other sentient living things?"

"Living things! I thought you only wanted me to name animals."

"And all sentient living things are animals."

"Is a butterfly an animal?"

"Yes."

"A grasshopper or a fish?"

"They are also animals."

"Strange! I thought grasshoppers were *insects*, and fish, I don't know what to call them."

"And *insects* and *fish* are sentient living things and therefore animals."

"To confess the truth" interposed Vidyabhushan, "all of us had a notion that quadrupeds only were called animals."

"I suspected so, but," said Mr. W. sympathetically, "you are no more guilty of such popular errors than many others who ought to know better. To tell you the candid truth, I myself formerly used the term 'animal' to designate the quadruped only; in fact I never took the trouble to enquire about it.

"I am waiting for an answer to my question."

"What is that, may I ask?"

"Forgotten so soon, eh!" My question was if you could see any animal here that you are already familiar with or others that resemble them.

"There are lots."

"Name some please."

"Here are the butterflies."

"Here are the crabs."

"What gigantic prawns!"

"Fancy exhibiting these wretched mosquitoes here as objects of curiosity."

"They *are* objects of very great curiosity indeed."

"How?"

"Just look at that enlarged drawing of a mosquito with its mouth parts modified for piercing and sucking. It has only one pair of fully developed wings, which are known as forewings; the hindwings are reduced to small knobs. Perhaps every day of our life we have killed many and disabled more of these little stinging pests, yet never have we paused to examine their organization and habits."

"I don't like them though, they are so blood-thirsty."

"Blood-thirsty no doubt, but with the majority of them the thirst for blood ever remains unsatisfied."

"We do not understand you. Blood, we thought, is as much necessary to a mosquito's existence as food and drink are to us."

"To be able to suck blood is no doubt a great delight

to a mosquito; but it is not necessary to its existence: in fact, the vast majority of mosquitoes never know what the taste of blood is."

"Indeed!"

"In all parts of the world there must be low-lying tracts of countries literally swarming with mosquitoes, where animals do not exist, or are extremely rare. Speaking of their blood-thirsty habit, a clever naturalist has suggested in respect to the mosquitoes of many districts of South America that perhaps only one in a hundred millions of mosquitoes can ever taste blood."

"How interesting."

"Why are these beautiful butterflies placed near the detestable blood-sucking mosquitoes?"

"Because they are near relations."

"Butterflies, near relations of mosquitoes! What next!!"

"I assure you they are. You can satisfy yourself whether it is so or not. Just look at these enlarged pictures and the dissection of the several parts of butterflies and mosquitoes, and prepare a tabular statement of the parts you see, in your note-book under the following heading, and say if I am right."

BUTTERFLIES.	MOSQUITOES AND FLEAS.
1. Body divided into three regions.	1. Body divided into three regions.
(a) Head or front region.	(a) Head or front region.
(b) Middle or chest region.	(b) Middle or chest region.
(c) Belly or back region.	(c) Belly or back region.

BUTTERFLIES.	MOSQUITOES AND FLEAS.
2. Limbs are built up of a succession of different pieces of movably jointed parts.	2. Limbs are built up of a succession of different pieces of movably jointed parts.
3. Mouth consisting of a spiral trunk forming a sucking tube.	3. Mouth consisting of a spiral trunk forming a sucking tube.
4. Both pairs of wings developed.	4. Only the fore-wings are developed.
5. The two wire-like feelers on the head are club-shaped and very prominent and many-jointed.	5. The feelers are small, not club-shaped and do not possess so many joints.

"We thoroughly understand it now. The butterfly and mosquito *are* closely allied to each other."

"But, my dear Mr. W., am I to understand that the cockroaches also are related to the butterflies? Because I see from this drawing and these dissected parts of a cockroach that its body and limbs are formed on the same model as those of a butterfly and a mosquito." The above question was asked by Vidyabhushan, whose countenance wore an aspect of great seriousness not unmixed with bewilderment.

"You are right, my venerable friend. All these creatures in which we find three distinct regions of the body, as we have seen in a butterfly, with three pairs of legs borne on the middle region, one pair of wire-like feelers, and two pairs of wings, belong to one class and are known as *Insects*. If further we carefully look at each region of these insects, we shall find that they are made up of a repetition of rings or segments." "What troubles you, my lad? Something appears to be weighing upon your mind. Speak out please."

"Did you not tell us," stammered Charu, to whom the above question was addressed, and who stood gravely stroking his beardless chin, "that the bodies of all insects are made up of rings or segments."

"I admit I said so. Am I wrong?"

"Oh no. I don't mean that. What I mean is that I have here found certain other animals, the bodies of which also appear to be made up of segments."

"Indeed! show us, please."

"Here are some prawn-like creatures, and there, some spiders too, whose bodies have segments."

I am very much pleased with you. It clearly shows that you have that spirit of enquiry which is so essential to the investigation of truth. You are perfectly right; the prawn and the spider, and, for that matter, the crab, scorpions and many others like them have also segmented bodies, and are therefore related to the insects. But, then, if you closely examine and compare a prawn, or a spider, with either a butterfly or a mosquito or a grasshopper, you will at once see that, although all of them have segmented bodies, the two former differ greatly from each other and from the three latter, in other respects. Just take a butterfly and a prawn as objects of comparison, and prepare a rough tabular statement of the parts you notice in each.

BUTTERFLY.	PRAWN.
1. Body composed of segments and divided into three distinct regions.	1. Body composed of segments: regions not distinct.

BUTTERFLY.	PRAWN.
2. One pair of feelers.	2. Two pairs of feelers.
3. Two pairs of wings.	3. No wings.
4. Three pairs of limbs.	4. Five pairs of limbs.
5. Body not enclosed in a shell.	5. Body with all the limbs enclosed in a shell.

Good. That will do for our present purpose. You see they are alike in some respects, yet unlike in others. In the fundamental plan of their construction, all these creatures, the butterfly, mosquito, prawn, crab, spider, scorpion, centipede &c., are alike, in that they all possess segmented bodies provided with jointed limbs. Taking advantage of this fact, Zoologists have given them a common name, *Arthropoda*, meaning animals with jointed limbs. But then, we have also seen that, although possessing these common characteristics, they are unlike one another in many other respects. Naturalists have again come to our help, and divided them into classes, orders, families, genera, and species, by grouping together such of them as have the greatest resemblance to one another. Speaking of this grouping we have already learnt that the animal Kingdom is primarily divided into two main sub-divisions :—animals that have bones or skeleton, and animals that have no bones. Among the latter we have seen many strange forms which we had never seen before, and about which we had but vague ideas, such as the Medusæ, the Zoophytes &c. Then we have seen that a large number of familiar creatures belong to this sub-division. You all know what a butterfly is, where prawns are found, and how toothsome, at least to some of you, they are. As for the mosquito, who among us can declare that he has never been a victim to its bite? In fact, you have known these things from your childhood, and yet, (I dare say you feel ashamed at your apathy), you never cared to enquire about them. But I am digressing. We have now studied the structures of some of these creatures and formed some idea of their position in nature, and the relation they stand in to one another. In short we have got some sort of a clue to their classification which may be roughly expressed as follows:—

A large number of animals, among which many are quite familiar to us, have segmented bodies provided with jointed limbs, and are therefore known in Zoological language as *Arthropoda*. All segmented and jointed-limbed animals may be divided into four great classes.

{

Class I. Prawns, crabs, and other animals like them, have two pairs of feelers; more than eight limbs; and the body enclosed in a shell or crust; and are therefore called *crustaceans*. There are other characteristics, but for our purpose these will be enough.

Class II. Spiders, and Scorpions, and animals like them have eight legs but no feelers.

Class III. Centipedes, and millepedes, are animals with numerous legs and one pair of feelers.

Class IV. Butterflies, mosquitoes, cockroaches, grasshoppers, houseflies &c., belong to this class. They are called insects; and all of them have one pair of feelers, three pairs of legs, and two pairs of wings, more or less developed.

"I have been connected," said Vidyabhushan, "in one capacity or another with the education of children and young men for the last thirty years, and have read and taught a great many things about animals and their ways as related in story and reading books. I know, as every school-boy knows, that lions and tigers are formidable animals; that ostriches are very large birds that live in the deserts of Africa, and are remarkable for their speed; that elephants are very sensible and asses very stupid. All these are very interesting and amusing to children, and have their uses. But then, this is learning things without method, and is, therefore, of no value. I am so glad that Mr. W. has hit upon this plan of teaching the teachers to value system. In fact he has given us a second sight. When I first entered this great hall, I was perfectly bewildered at the vastness of the collection, and had not the least idea in what order and plan they were arranged. I have got at least some notion now of their arrangement, thanks to the interesting demonstrations of Mr. W. We ought to be very grateful to him."

"Since you so much appreciate my system, let us complete our knowledge of *the animals without bones*. It is, however, already past four, and we have to attend the evening lecture of Professor ——, so we had better postpone further consideration of the subject till to-morrow."

POND HUNTING.

Talking about the various interesting experiences of the day, Mr. W's. party were proceeding to-wards the esplanade. But where is Mr. W.? Gone to his club? Oh no, that is impossible; he is to accompany his pupils as far as the nearest cabstand. Good heavens! there he is!! *sans* shoes, *sans* socks, with trousers tucked up at the knees, he is searching for something at the water's edge in yonder tank in the maidan. Gentle reader! your sense of propriety need not be shocked at this undignified appearance of an Englishman, and a high official to boot. Ask Vidyabhushan or any of the elder Pandits, and they will tell you all about his pursuits and predilections; how, when out inspecting schools, they have often seen him deliberately climbing some roadside tree to look at bird's eggs, or spending hours in watching butterflies disporting themselves in the air. It was not long ago that he was found carrying a load of rare plants which he had chanced to find in some out of the way place. Vidyabhushan once asked him why he took so much trouble about such trifles. What do you think was his answer? It was characteristic of him. He said that there was nothing trifling in Nature, and, as for trouble, he loved constantly to refer to his great mistress, meaning Nature, for knowledge and experience. Let us observe what he has been doing now. He is dipping up dirty water from the tank. Just take note of the appliances. There is a small net made of coarse muslin, its top part is sewn on to a steel ring about six inches in diameter; attached to the ring is a ferrule which nicely fits into the end of Mr. W's walking stick. The bottom part of the net, which is about two inches in diameter, is tied on to the neck of an ordinary four ounce round phial. Look, wherever there is a gleam of sunshine upon the surface of the water, he is plunging his tiny net every now and then, and raising it when withdrawn slightly above the surface so as to allow the waste water to flow through the net. Having got a haul, he transfers the contents of the phial into a larger vessel.

"I wonder why he chooses sunny spots."

"Because he knows that the water animalcules are sure to come to the surface there."

"What are you doing, Mr. W. ?"

"O, is it you, Vidyabhushan ? I am pond-hunting for to-morrow's lessons."

HALF AN HOUR WITH A MICROSCOPE.
(Second day in the Museum.)

Arranged on a long and narrow table in the spacious verandah of the Invertebrate Gallery are two or three microscopes, a glass containing some dirty water, a few glass tubes with tapering ends, and one or two glass rods. Mr. W. takes hold of one of the glass tubes between the thumb and the third and fourth fingers of his right hand; the index finger is pressed firmly on the top. The tube is dipped into the water of the jar, and its open end placed over the sediments which have collected at its bottom : the index finger is now slightly lifted, causing a gentle rush of water into the tube carrying some sediments ; the finger is immediately applied to the top and the tube withdrawn. The contents of the tube are received into a watch-glass and examined with a lens. There is something in it, and this must be properly examined under a higher power. By the aid of one of the glass rods, a drop is transferred to a glass-slide and placed on the stage of the microscope.

"Do you see anything?"

"Nothing that we can make out, sir."

"Focus the microscope properly, and look again."

"Ah yes, there is something gliding across the field of the microscope. How funny."

"Funny indeed! But let us hear what they look like."

"They look like small masses of some jelly-like objects of oval or rounded forms."

"You have not seen them Hari Charan, neither you nor my friend Vidyabhushan."

"Did I not hear you describing them as of oval or rounded forms? As far as I can see they are very irregular in shape, all bays and promontories, that is, projections on one side and depressions on another.

"So they are."

"Do you know what they are?"

"No sir, we don't."

"From the fact of their movements you must have concluded that they are no mere masses of dead matter but something endowed with life, and so they really are. They are the *Simplest forms of Living things* and the Zoologists call them *Amœba*."

AMŒBA.

They are mostly invisible to the naked eye, as they rarely exceed one hundredth of an inch in diameter. The most noticeable thing about an Amœba is, that it is never quite the same shape for long together. With regard to this change of form, a distinguished Biologist says that it takes place so slowly as to be almost imperceptible, "like the movements of the hour-hand of a watch, but by examining it at successive intervals, the alteration becomes perfectly obvious, and at the end of half an hour it probably has altered so much as to be hardly like the same thing." Amœbas are composed of a more or less jelly-like substance called "protoplasm," which is semifluid in character and is a very complex chemical compound of Nitrogen, Carbon, Oxygen, Hydrogen, and Sulphur &c. They are the simplest forms of living things, each consisting of a single particle of protoplasm which performs all the various functions by which life is maintained. Dr. Hudson describes them as "slow gliding lumps of jelly that thrust a shapeless hand out where they will, and, grasping their prey with these chance limbs, wrap themselves round their food to get a meal; for they creep without feet, seize without hands, eat without mouths, and digest without stomachs."

PROTOZOA.

There are numerous other kinds of animals, or, more correctly speaking, animalcules, consisting of single particles of Protoplasm, which are all collectively called

Protozoa. They are the most lowly organised members of the animal kingdom, or *the Beginnings of Life.*

" Do you know what that thing is, Charu ?"

" Chalk, I suppose."

" Well, you are right so far that it is the white chalk of Geology. Let me tell you what it is and how it is formed. There is a kind of aquatic *Protozoa* called *Foraminifera*, that is animalcules with perforated cell-walls; their tiny body is enclosed in a tiny shell which is usually perforated with little holes like a sieve. They are mostly marine animalcules, and countless numbers of them exist at the surface of the sea; as they die, their shells fall to the bottom where they form deposits of great extent. Those of you who are interested in this branch of knowledge and wish to know something further about the single-celled animals or animalcules must read up some works on Systematic Zoology."

MANY-CELLED ANIMALS.

" All the rest of the animal creation, from a sponge or a worm to a man, are comprised under this group; but before we proceed to inspect the various forms of animals other than those we have already seen, I should like to impress upon your minds a few simple, yet important facts in regard to these *many-celled animals*, and I think I had better do so in the language of a competent Zoologist."

' Among the *Protozoa* the single simple cell perfor ms,

though imperfectly, all the various functions by which life is maintained: it receives impressions from without, *e. g.* of the proximity of food, it is locomotive, it takes hold of and digests prey, it respires, it circulates its assimilated nutriment, and it excretes its useless waste,— and in most "*Protozoa*," as you must have noticed in *Amœbas*, "it does all this with any and every part of its simple unicellular body.

'But when a multitude of cells is incorporated together to form a large many-celled animal, it is not economical for the incorporation that each of its cell-units should, so to speak, go its own way and continue to perform all its functions independently for itself. It is more economical, both for the incorporation as a whole, and each of these cell-units, that each of the functions requisite for the common weal should be assigned to particular groups of cells—to some group the receiving of impressions, to other groups the functions of digestion and assimilation, and so on : much as happens in communities of men, where the various industries and activities of the community are not performed by all the citizens alike, but each citizen is devoted to one particular occupation. In this way, since 'practice makes perfect' each function comes to be dexterously and economically performed.

'This apportioning of functions is known as the "physiological division of labour," and we must now consider the changes to which it gives rise.

'Let us imagine a colony of cells, all of which are

alike in form, and, like a Protozoon, possesses the same limited range of ill-developed functions, and let us now suppose that to one group of cells is allotted the function of protecting the colony from external injuries, and to another group of cells the function of digesting the food of the colony. Then as a natural result of special perfection in one set of protecting functions to the relative neglect of the digestive functions, and *vice versa*, these two groups of cells will gradually come to differ in many points of structure and form.

'In other words, among cells originally alike in form the specialization of function to which the division of physiological labour gives rise, leads to many differences in form.

'A second consequence of physiological division of labour, and of the differences in occupation and structure to which it leads, is that all the various cells and groups, like the citizens of a state, come to be mutually dependent on one another.'*

"Having heard so much about the many-celled animals and of the "physiological division of labour" in them, you must be naturally impatient to see some such types in which this physiological division of labour can be seen in its earlier and simpler stages. Well then, here are our old friends the *Zoophytes*. Unlike the *Protozoa* in which the single unit of protoplasm performs all the functions of life, the constituent cells of the body

---

* Guide to the Invertebrate Gallery of the Indian Museum, by Surgeon—Captain A. Alcock, Superintendent of the Indian Museum.

of a Zoophyte or *plant animal* fall into two distinct layers of membranes—an external layer whose business is to protect the animal and recieve impressions, and an internal layer for the purpose of digesting food."

SPONGES, CORALS &c.

" Now let us look at these cases which contain a large number of strange, beautiful and grotesque forms. All of them, as will appear from the labels, are *many-celled animals*, and are collectively called *Cœlenterata*, that is, animals in which the cavity of the body is formed entirely by the intestine. Examine these animals somewhat closely. What do we find? Here are *sponges* and *corals*, objects with which all of you must be familiar; then, there are the *sea firs*, the *sea jellies* and the *sea anemones*, which your more fortunate brethren of the Madras and Bombay coasts must know better. As a boy, I had an aquariam full of these interesting objects.

" If all these animals are carefully examined, some of them will be found to differ very characteristically from the others; hence the Zoologists have divided them into two principal types. Now, here is an opportunity for you to exercise your powers of observation. Tell me, please, after having inspected the specimens minutely, if anything peculiar strikes you? Yes, you are perfectly right, some of them have neither feelers nor spines, whereas, others *are* provided with feelers and spines. The former belong to

"I. *the sponge type*, and may he defined as jelly-like organisms 'destitute of a mouth, and united into a composite mass, which is traversed by canals opening on the surface, and is almost always supported by a frame-work of horny fibres, or of silicious and calcareous spicula.' It is only the skeleton washed and dried that you see exhibited here. By the by, have any of you ever seen a live sponge? Why look surprised? Oh! that is your idea. Let me assure you then that sponges are not only found in the sea but in fresh water also, and that the *spongella*, as this fresh water family of sponges is called, is found in marshes and stagnant waters in every part of India. Do you know where the sponges come from? Yes, the sponges of commerce come from the Bahama Islands, and from the Grecian Archipelago also. But then there are other kinds which are found in deep seas. Didn't one of you ask me why sponges are labelled *Porifera*? Let us hear what Upen has to say, I know he has been working hard at Latin to qualify himself for the Gilchrist examination. You are very right, *porus* in Latin means a pore; and *fero* I carry, and the name has reference to the fact that in sponges respiration and ingestion of food are carried through the pores in their body. Before we say good-bye to the sponges, I wish you to take note of the various kinds of sponge skeletons here displayed. We now come to the Zoophytes with feelers and spines which belong to

"II. the *Stinging Zoophyte* type. The stinging cells or spines as you call them constitute their weapons of offence and defence. Although not as easily seen as the tentacles and spines, yet if one of these animals be carefully examined, a mouth will be sure to be found through which food is ingested. The simplest form of a stinging Zoophyte is seen in a

"HYDRA, an animal that can only be properly examined under a microscope. But we need not despair of seeing what it is like, as it is here represented in the form of an enlarged drawing from which it will appear that it is an animal having the form of a simple tube, closed and often attached at one end, and provided at the other with a mouth surrounded by tentacles. The hollow of the tube forms the gastric cavity or stomach.

"There are other and more complicated forms of stinging Zoophytes, as a glance at these cases will show. Here is for instance a tree-like colony of animals. If we were to trace the life-history of this colonial form, we would probably find that some single individual of the Zoophyte class budded and branched again and again; and that the resulting offspring instead of separating themselves to live individual lives all remained united together and to the parent, to form this tree-like colony. As, however, it would be impossible, within the limited time at our disposal, to study in detail all the various forms of stinging Zoophytes and the interesting gradations of their structural peculiarities, we must for the present, content ourselves with simply inspecting our old friends the *Medusæ* and the *Corals*, as instances of further advance in structure and in the physiological division of labour.

"Some idea of the gigantic results of this division of labour may be formed, if I tell you that large Islands are formed by the labours of certain kinds of stinging Zoophyte known as Corals. Not to go far, the Maldive and Laccadive Archipelagoes are good examples of such Coral Islands. You want to know how they were formed. It is simple enough. In ages past millions upon millions of Corals gave rise by budding and fission to great colonies. These having come together in shallower waters formed vast reefs: so vast that they ultimately became raised to the level of the sea and laid the foundation as it were, of the present groups of Islands."

WORM TO MAN.

"We now come to altogether a different series of animals, which are as interesting as those we have hitherto examined. In the latter we did not find any trace of a posterior or tail end properly so called. Nor by any stretch of imagination could the body of any of those animals be said to possess ventral or dorsal aspects; that is, a lower surface turned towards the ground, and upon which the animal moves, and an upper or back surface. The animals of the present series seem, however, to have been constructed upon a different plan, as, in the majority of them at least, we can distinguish a head and a tail, and also an upper and a lower surface. Further, all the animals of this series are characterised by the possession of a true body cavity, through which the alimentary canal runs, somewhat, as a Zoologist of repute has graphically described, like a lamp-chimney through a lamp globe. Besides the alimentary canal, the organs of circulation, excretion, and reproduction are also lodged in this cavity.

"We have no time however to study in detail all the various types of animals of this series; suffice it to say that their number is very large and that their forms are various with an immense diversity of gradations—from worm to man.

"*Lower type of worms.* Do you see those tape-worms there? They are the most degraded animals of this series. Degraded, because they are incapable of leading an independent existence, living commonly as parasites in the alimentary canal of warm-blooded vertebrate animals: they are also structurally incomplete, the alimentary canal being almost wanting or imperfect. Each of them has a wonderful story to tell you about the complicated series of changes it went through in the course of its development from the egg to its present mature form. The egg after having escaped from the sexually mature segments of the parent tape-worm found its way into the stomach of some warm-blooded vertebrate animal, say a cow. Here it did not at once develop into a tape-worm complete in structure like one of those in that case before you, but elected to assume the form of an embryo provided with hooklets. By means of these hooklets it eventually bored its way through the stomach of its first or intermediate host, as that cow must be called, and escaped into one of the blood-vessels, and was carried to some distant part of the body. Here it remained embedded in the tissues, and even gave rise by budding to other embryos like itself; but attained no further development. In the meantime the cow was eaten up by another vertebrate animal, say a tiger, and the tape-worm embryo having reached the alimentary canal of its second or tiger host developed itself into fully mature tape-worms as you see them before you.

"Ah, I see Vidyabhushan smiles. Do you understand why? In this short life history of a tape-worm he finds additional argument in favour of his vegetarian principle. Who knows if the animal we eat is not an unconscious host of a tape-worm guest?

"*Higher types of worms* comprise the round worms and thread worms. They are on a higher level than the *tapeworms* because they have a distinct body cavity, and their reproduction is sexual.

Higher grade of animals with true body-cavity. "Animals of the next series are structurally more perfect than the tape-worms and round-worms. Hari Charan says some of them look like common earth-worms and leeches. They cannot help doing so, I am sorry to say, because they are as really earth-worms and leeches as you and I are men. Just try and describe the external appearance of one of them. Very good. I must say you are very clever. You have hit upon the right points:—(i) a body made up of a series of equal rings or segments; (ii) an anterior or head segment,—why omit to mention that, behind the head segment is the mouth; (iii) a posterior or anal segment in which is the anus; (iv) the upper and lower surfaces of the body, that is, the dorsal and ventral surfaces as they are called in Zoological language. Allow me to supplement your excellent description by stating that if any of these creatures be divided lengthwise through the middle of the body, you get two halves of equal symmetry.

"There are other details of their structure which we had better postpone the consideration of, until we have made some progress in our study. It would suffice for our present purpose, if I simply call your attention to the fact that there are a great many other animals belonging to this series. Here, for instance, are some sea-worms from the Andaman reefs; they are provided with a distinct head, powerful jaws and well-developed locomotive appendages. Do you see those curious-looking objects next to them? They also are sea-worms; but as they live a different life, they are differently equipped. Unlike their brethren of the Andaman reefs which lead a free swimming life, these are sedentary in habits, and live in protective tubes or shells which they either manufacture or secrete for themselves. Here then are two good examples showing how the manner of living may modify structure.

"The next series comprises animals with segmented bodies and jointed limbs which the Zoologists have designated *arthropoda*. But as we have already considered them somewhat in detail, let us proceed with the inspection of a most singular group of animals exhibited in these wall cases. As most of the animals of this group have spiny skins, the Zoologists, who delight, in the same manner as Indian Pundits like our Vidyabhushan, in high sounding names, call them."

ECHINODERMATA.

"The name is no doubt very suggestive of the peculiar outward characteristic of all these animals, as in simple Queen's English it means an animal with hedgehog-like skin, from (*Echinos*, a hedgehog which has a spiny body, and *derma* skin).

"They are marine animals usually of the shape of a five-rayed star and covered with a skin which is always thick and stout and is often so impregnated with lime salts as to form a rigid stony case. For further protection, the hard stony integument is often thickly covered with short spines.

"The animals of this group are of a higher organizaion than any of those we have hitherto seen and studied. The proof of this lies in the following facts :—(*a*) They possess blood-vessels in which colourless blood circulates; (*b*) they have a nervous system ; and in the starfishes (*c*) simple eyes are placed at the tips of the rays, these eyes being so small that it is difficult to trace them.

"A look at the animals displayed in these cases will show how varied are their forms. Some have a flat central disc with five regular flat pointed rays. On the under surface of the disc at its centre is the large mouth. There are others which have so many as thirteen or even twenty-one rays instead of five. The animals are called *Starfishes*, which are found at all depths of the sea, from the tide-mark down to three-thousand fathoms deep. What do you think these animals live upon? The shallow-water species live upon the *mollusks* and the smaller *crustacians*, which they find in abundance in rocks, reefs, and beds of hard sand, which form their congenial habitation. The deep-water ones gorge themselves with the ooze of the ocean bottom, where they have their habitation, for the sake of minute particles of living or dead animal matter.

"There are a great many other kinds of Echinodermatous animals all named in reference to their forms and character. The *Brittle Starfishes* which flourish best in the tropical seas ; the *Sea Lillies* and *Feather-stars*, here represented, are all from the Indian seas. The *Sea-urchins* and *Sea-eggs* are found in all seas and at all depths. The *Sea-cucumbers* are also found in all seas and at all depths. The Chinese are said to be very fond of Sea-cucumbers, which they spend much time and trouble in procuring on festive occasions."

THE MOLLUSCA.

OR

SOFT-BODIED ANIMALS.

"The Molluscs constitute the last group of the *Invertebrata* or animals without a skeleton. A glance at the contents of the cases labelled "Mollusca," will show what a great variety of forms they must have, and how grotesque-looking some of them are. In some of the groups hitherto examined, we found one or more types such as were quite familiar to us; for instance, among the vast multitudes of animals known as *insects*, we came across the *butterflies*, the *cockroaches*, and the *grasshoppers*; then, there were the *leeches* and the *earthworms* among another group, and so on. Now, let us carefully examine the Mollusca group, and see if we can discover any familiar forms of animals among them. It is exactly as I anticipated. Just look at these creatures, please, and tell me what they are? Quite right, they are snails, and there are a great variety of them, including our common garden and pond snails, (*gendi, samuk, gugli*). The garden-snail may be regarded as a typical Mollusc, with a soft body, without segments, rings, or rays, protected by a hard shell, into which the body can be withdrawn. All of you must have seen garden-snails at one time or another, and I dare say could not have failed to notice that it has a distinct head with a mouth, and a pair of short broad "lips," which in Zoological language are called "mouth tentacles;" also that it has a pair of long wire-like head tentacles, upon some part of which the eyes are placed. Have you ever attempted touching one of these snails while slowly and deliberately crawling along? What has been the result? It quickly retires into its house or shell. Here is an opportunity for you to study the other parts of a snail from this dissected specimen. The body is enveloped in a fold of skin which is called its "mantle;" the under part of the body is thickened to form an elongated solid foot for progression; the thin-walled upper part contains the viscera and is permanently placed within the cavity of the shell. The internal organs of a typical mollusc consist of (1) an alimentary canal, coiled and embedded in a dark digestive gland or liver; (2) a two-chambered heart; (3) an excretory organ or kidney; (4) the reproductive gland; (5) the nerve-ganglia placed in front of the alimentary canal; (6) the salivary gland. Another characteristic organ of a typical mollusc is the "tongue," or "odontophore," as it is called by the Zoologists. It is a horny strap, of which the upper surface is generally covered with small sharp teeth-like projections arranged in rows like a rasp. When not used, it lies coiled up on the floor of the mouth, but when in use it acts backwards and forwards like a saw."

OTHER FAMILIAR MOLLUSCA.

"I shall next call your attention to some very interesting objects, the *conch* and the *cowries*, which must be quite familiar to every one of you. Yet I doubt whether you ever troubled yourselves to enquire about their history. My friend Vidyabhushan is candid enough to confess that he has never done so, although a conch has formed a part of the paraphernalia of his daily worship for the last forty-five years. Let me, however, tell you something about these *molluscans*.

"The conch (*sankha*) are marine shells, which vary in shape and size according as they belong to one family or another. The animals inhabiting the shells are all carnivorous, and some, you will be surprised to hear, are even carrion feeders. The common *shank shells* (*Turbinella pyrum*) are abundant on the coasts of India, Ceylon, Siam, and China. "Pap boat" is the name of another kind of shell common on the coast of India and Ceylon. Writing about this shell, Sir Emerson Tennent, a former Governor of Ceylon and a great Naturalist, says that, it is used by the Hindus of the Malabar Coast to keep sacred oil which is employed in anointing their priests.

"The "*Spinde shell*," called "Buckie" in Scotland, is extensively used as an article of food by the poorer classes of people there. By the by, have you ever heard of the "Roaring Buckie," in which the sound of the sea may always be heard? Who is whispering those lines there behind me, ? They sound like music to my ears. Go on, please, speak aloud.

> "I have seen
> A curious child, who dwelt upon a tract
> Of inland ground, applying to his ear
> The convolutions of a smooth-lipped shell.
> To which, in silence hushed, his very soul
> Listened intensely ; and his countenance soon
> Brightened with joy ; for from within were heard
> Murmurings whereby the monitor expressed
> Mysterious union with its native sea.
> Even such a shell the universe itself
> Is to the ear of Faith ; and there are times
> I doubt not, when to you it doth impart
> Authentic tidings of invisible things ;
> Of ebb and flow, and ever-during power ;
> And central peace, subsisting at the heart
> Of endless agitation. Here you stand,
> Adore and worship when you know it not ;
> Pious beyond the intention of your thought ;
> Devout above the meaning of your will."

The shell alluded to in the above lines, is really used by children, in certain parts of England and Scotland, in the manner described by Wordsworth. But, my friends, it is not merely for the sake of the allusion that I asked Upen to recite the above exquisite lines. My object was to take this opportunity to impress upon your minds, that, in order to understand Nature, and to be able to receive her teaching, we must have certain qualities, and these, according to the philosophy of the poet quoted, are the "qualities of the child." The first and foremost is a simple heart, as that of the child whose countenance brightened with joy when he heard the murmurings within the convolutions of a shell. Faith and purity of heart are other essential qualities of a child. Any one who cares to enter the portal of knowledge must do so with a loving and listening heart, and with the ear of faith.

"One moment now may give us more
Than years of toiling reason ;
Our minds shall drink at every pore
The spirit of the season."

"I thank you Mr. W.," interrupted Vidyabhushan, "from the very bottom of my heart for your interesting and instructive discourse upon the qualities of the child. I wish we were all as simple-hearted and loving creatures to-day as we were in those blissful days of our childhood, when the shadows of worldliness had not darkened our souls. But what I wanted to tell you is that according to our old Hindu idea, 'Reverence' is another essential quality for the training of the mind."

"You are perfectly right, my friend Vidyabhushan, and I am much beholden to you for your suggestion, but I thought it superfluous to preach 'Reverence' to a party of educated Indians, as, whatever their other faults may be, the charge of irreverence cannot be laid at their door."

"I sincerely wish you were absolutely right," replied Vidyabhushan. "But," quoth he. "I have noticed with grief and concern a spirit of irreverence growing among our educated young men. Whatever the causes may be, the evil is there, and, unless it is nipped in the bud, it will grow like an Upas tree to poison the atmosphere far and wide. How well do I remember the day when my wise and affectionate father initiated me into the mysteries of the Bengali alphabet, and taught me the following sloka—

एकमप्यचरं यं तु गुरु: शिष्यं प्रबोधयेत् ।
पृथिव्यां नास्ति तद् द्रव्यं यद्दत्त्वा सोऽनृणी भवेत् ॥

Ever afterwards, he lost no opportunity to teach me by precept and example, to love and fear God, and to bend in loving reverence before the "beauty and majesty of the universe" and to be always respectful to my teachers and superiors. Such, however, has been the wide-spread degeneration of the present day that even parents do not take the trouble to train up the minds of their children, but are content with leaving them to their books and teachers. But I am digressing, and I must ask your forgiveness, Mr. W., for causing this interruption to your interesting discourse."

"On the contrary," replied Mr. W., "you deserve my best thanks for having incidentally called my attention to the important subject of juvenile training. In fact, every one connected with the problem of education in India ought to ponder over the subject."

THE COWRY SHELL.

"I must call your attention, before we leave this hall, to another class of familiar molluscs the *cowries*. They are largely used as currency among the poorer classes of Indians, and are procured in vast quantities on the coast of the Congo and in the Philippine and

Maldive Islands. As a medium of exchange, the Cowry must have been introduced at a remote past, when civilization was in a rudimentary stage. It is still used as currency over an immense area extending from the south Asiatic countries to the interior of the African continent. In this family of shells, the head of the animal is large and broad, the tentacles or feelers long, and the eyes placed on projections on their external bases; the foot is large, expanded, and elongated behind.

The cowries are nearly all tropical animals, inhabiting Coral reefs and submerged rocks. In their habits they are shy and slow.

VII.
EXTRACTS FROM AN ANONYMOUS JOURNAL (II).

THE PARK.

A delightful morning! There was a serene freshness everywhere. The dew-washed foliage and grass looked green and lovely. The air resounded with the wild minstrelsy of innumerable birds, and every tree and every bush had a colony of its own. From the graceful *kadamba* and sombre mango came rich delicious strains such as *doyal* birds only can sing. *Shaliks* and sparrows and little sun-birds chirped on trees, on shrubs, on hedgerows and housetops. Conspicuous, because most demonstrative, were the crows which were busy cawing and fluttering on every side. Far off the air was full of a wailing cry, harsher and more piercing it sounded as nearer and nearer it came ; it was a solitary kingfisher calling to its mate as it winged its way towards the distant marsh. Already there were vultures and kites soaring very high, and appearing like tiny specks against the azure sky. The sun was just rising, and the eastern sky looked glowing and glorious. In a moment the tall tree tops became bathed in soft golden rays which gradually penetrated into the poor man's hamlet and the rich man's hall, and illumined everything. The birds had done singing their morning anthem, and now hopped about and played in all directions, infecting us with their joyousness. In a merry mood congenial to this changing and charming aspect of Nature, we wended our way, one fine morning, now several years ago, towards the open fields and meadows.

We had almost reached the outskirts of the town and were about to leave the dusty public thoroughfare and strike eastwards towards the park, when, we observed a series of interrupted double lines right across the road. They looked like foot prints of some bird, but there were no marks of claws or nails. Our curiosity and interest having been aroused we began a closer examination of them and their situation. They all converged round the scattered remnants of a heap of horse-dropping on one side, and were lost in the grass and shrubs bordering the road on the other. We tried to trace the marks into the shrubbery with a view to finding the cause, but found it difficult, owing to the sodden state of the grounds, and abundance of fallen leaves and twigs decaying and decayed. We routed about among the grass and *debris* for some distance and were about to give up the search when suddenly something tiny was noticed disappearing in a thick tuft of grass. It was a snail! One of those common garden-snails with a shell on its back. There was another, and its position in regard to one of the series of marks unmistakably pointed it out as its cause. In order, however, to satisfy ourselves on this point, we placed the snails on the middle of the road, and retired to a short distance; in a little while they began crawling away leaving behind them trails similar to those we had observed. There was no doubt now that these molluscs had crawled out of their hiding places in the thickets to feast upon the heap of droppings, which with other vegetable matters constitute their principal food. We put the snails into our collecting basket in order to examine their structure and study their natural history at leisure, and resumed our walk.

For a short distance our route lay through pretty groves of mango, guava, leeche, and other fruit trees, beyond which lay extensive corn-fields, now carpeted with a profusion of healthy green plants. Following one of the raised tracks separating adjoining plots we reached the foot of an embankment. Its sloping sides were covered with impenetrable thickets of *sheakul*, *apang*, Indian nettle, and other herbaceous plants. A slight ascent to its brow, a few minute's walk along its narrow top, an easy descent, and again a few minute's walk by a shady winding footpath and we reached our destination. It was a lovely little park-like place, well stocked with noble trees and nobler palms, growths of a century perhaps. A wicket gate in a gap of a thick *mendhi* (Indian myrtle) hedge gave us admission into this woodland scenery. An avenue of *bakul* and *piyal* led us to an open space, in the centre of which stood a cottage, a real *kūtir*. Built of dab and wattle and thatch it was rough, rude, and rustic, but a model of cleanliness, cosy and comfortable. Creepers of various kinds covered its sides and roof, and a mighty banyan gave it shade. There was a fine lawn in front of it and all round it were shrubberies and arbours. To the west and within a short distance of the cottage was a nice little pond of irregular shape—a lake in miniature, surrounded with groups of palms and pandanus, which formed an agreeable background to its gently sloping grassy bank. Of under-growths, there were numerous species of aroids and ferns, the latter forming clusters round the thick stocks of the palm trees. The pond was full of various kinds of aquatic plants but the most conspicuous was the *padma*, which lay idly but with resplendent beauty on the surface of its still water.

Here, away from the struggles and strifes which agitate the busy world, lived Yogananda Svami a true Vaishnavite, justly renowned and respected for his scholarship, piety, and devotion. At the time we speak of he was long past the prime of life, but was yet "light of foot," and erect in bearing. There was a happy mixture of benevolence and austerity in his countenance, and in him, the polish and elegance of a scholar and a gentleman united with the bluffness and severity of an anchorite. His sympathies were wide and catholic, so that, if there was any case of sickness anywhere within reach, he was sure to be there ministering to the wants of the patient, and to all such his very presence was a benediction. He was the invariable arbitrator in all family disputes, and there was hardly any who ever objected to cheerfully abide by his council. Gifted by nature with a superior intelligence and an acute power of observation he had improved both by study and travel, and it was always a pleasure to hear him talk about his experiences of the various parts of India.

Ever ready to take ignorance by the hand he would begin talking upon a subject, draw us into discussing the matter with him, and would let the moral come from the conversation naturally without making obtrusive efforts to force the truth into our minds. A distinguished graduate lately remarked to me that, in his youthful conceit he had imagined himself as a young man of finished education and culture, but a fortnight's intercourse with the good and gifted Svamiji undeceived him. He quickly came to perceive that while he had acquired a certain proficiency in what for a better expression may be termed "book-learning," his faculties of observation remained wholly undeveloped, and that he was shamefully ignorant of the commonest things that surrounded him on every side. For ourselves, we found Svamiji a most agreeable instructor, full of sweetness and light. He often took us round the park and talked to us about the trees and shrubs, and taught us the alphabets, as it were, of the plant-lore.

ON SOME TREES AND SHRUBS.

A stately *ashwathwa* (*Ficus religiosa*) is a noble tree wherever found. The one, however, to which reference is here made, and round which hover many associations is a particularly fine one. It stands on the south-west corner of Svamiji's park bordering a public path, and is surrounded by a spacious mud platform with brickwork facing. Morning, evening, or noon it is a place of much resort. During the fierce midday heat of summer weary pedestrians find rest under its cool and grateful shade. Simple village folks from many miles round come to worship the tree, or, more correctly speaking to offer *poojah* to the presiding gods of the tree. For, according to popular Hindu notion, Brahmá is supposed to hold sway over the roots of an *ashwathwa*, and Vishnu over the stem ; the branches and leaves being assigned to Siva with his retinue of ghosts and goblins. Small wonder that people should feel religious scruples in breaking even a twig of this sacred home of the Hindu trinity, or horror at the idea of being possessed of evil spirits which hold nightly revels under its shade ! Whatever the origin of the belief might have been, it served, according to Svamiji's idea, one very useful purpose—that of saving such valuable and slow-growing trees from destruction in a country where cool shade is a necessity during the greater part of the year. With the womenkind of the neighbourhood it is an object of especial veneration. One of the earliest impressions that dream-like flit across my mind's eye is the dim recollecttion of a large tree, a crowd of women and children dressed in their holiday's best, and a profusion of flowers, fruits, and eatables. It was a great festive occasion, when devout mothers from far and near were gathered together under the ample shade of the *ashwathwa* to worship *sasti*, and invoke the blessing and protection of the benign goddess upon their darling sons and daughters. This happened at that interesting period of early childhood when we begin to understand the difference between a cow and tree, a goat and a dog, when, in fact, we just commence learning the elements of science which, in reality, consists of the knowledge of one thing from another. Since then often have I witnessed the ceremony, in which mothers prayed and grandmothers gossiped, children frolicked and danced, laughed and sang.

The Ashwathwa is a living being. We have grown wiser, a new soul as it were is born in us, and we now look at the familiar yet venerable *ashwathwa* not in the old light, but as an *organised being, endowed with the properties of growth and reproduction*. As a living being it must needs have food and drink, both of which the *ashwathwa* has enough and to spare. In order, however, they be able to understand what the nature of the food and drink is, and how the tree takes them we must begin at the beginning.

Popularly speaking, our notion of an organised living being is that it must have limbs to move about, mouth to feed with, stomach to receive and digest food, eyes to see, ears to hear, lungs to breathe, and so on; or, some modifications of the appendages and organs mentioned above. But in the *ashwathwa* we do not see anything that can, even by a stretch of imagination, be called limbs, mouth, stomach, lungs &c., or any modification of them. Nevertheless the act of eating, drinking, breathing is incessantly going on there. How, we shall presently see. Our previous acquaintance, however slight, with a vast multitude of sentient living beings has taught us to see nothing strange or unusual in infinite modifications of structures in living organisms according to their modes of life and surrounding conditions. Extending our observations a little further, we find that the limbs which fit a cow, a goat, or a man for a terrestrial life, are modified into wings in birds, which fly in the air, and into fins in the fish, which swim in the water. This principle of modification of structure is carried further into effect in the *ashwathwa*. For though an organised living being in its highest development, the *ashwathwa* is incapable of voluntary motion. On the contrary it is firmly fixed to the ground, so firmly that it is capable of withstanding the force of wind and storm with impunity. Instead, therefore, of limbs, or, any modification of them, it is provided with strong and heavy anchors in the shape of large and long roots, which are many and penetrate deep into the soil. But to fix the tree to the ground upon which it grows is not the only function of the roots; *they also suck up the moisture from the soil.*

There is yet another very important function which the roots perform, they absorb *nitrogen, sulphur*, and *phosphorus* from the soil. It must not be understood, however, that they take in these elemental bodies in their free state as *nitrogen, sulphur*, and *phosphorus*, but in combination with other things in the form of soluble *nitrates, phosphates* and *sulphates*. Why this particular function of the roots is so important, and what part these elementary bodies, especially the *nitrogen*, play in the life-history of a plant, will be explained later on, when we come to study other parts of the tree.

Strange though it may sound it is nevertheless true, that in the performance of their several functions the roots might almost appear to exercise a certain amount of intelligence and discrimination. The evidence is close at hand. There, upon a tumbled-down brickwork—the fast disappearing remains of an ancient temple, grows a banyan. It is yet a small tree, but its cable-like roots are already very long, and rootlets numerous. The obvious impression which they convey to an observer's mind is that, inorder to fix the tree with firmness, they are struggling to reach the ground as quickly as possible, sending down guy-lines in all directions; whereas the rootlets are busy searching for holes and crevices to insinuate themselves into the soft soil. Let us look at them somewhat closely and narrowly. Here, a set of thread-like small roots have reached the hard surface of a brick, and having found out their mistake are retracing their steps as it were, and creeping down its sides in a network of loops; others have just touched the bricks and stones, felt their hardness and avoided them. On the other hand, some bolder spirits among them have marched right across the bricks and stones to reach the common destination, the mother earth. Some invisible yet loving hand helps and guides these tiny rootlets in their progress through dark holes and crevices, and in their search after the prime necessity of life—the water.

It is not only the roots, or the parts adjacent to them that drink the water, but every part of the tree, even the tiniest leaf, at the very top of the tree which in the case of the *ashwathwa* is at least forty to fifty feet above the level of the ground, receives an adequate quantity of the moisture which the roots suck up. The way in which the tree manages to accomplish the work of raising and distributing water to such a height is very curious indeed; and we have an interesting example of it in the water supply of large towns and cities. First, there is the system of pipes for the conveyance of water; then, there is the motive power supplied by complicated machines for forcing up water for distribution. What human skill and ingenuity have done for cities and towns, the Divine Author of the universe has done for the tree, providing the roots and stems with a perfect system of pipes, through which water is carried upward by virtue of some subtle force breathed into them by that Great Author. What is that *subtle force* that is employed in raising the water? We had better enter into a little detail.

How roots and stems convey water. The roots are made up of cells placed end to end; so that those at the very extremity of the root fibres and root hairs are necessarily in contact with the earth. These cells have a special power of absorbing moisture from the soil with which they are in content. They go on absorbing water and passing it onwards to neighbouring cells and the water is thus ultimately transferred to the water-conducting tissues of the wood in which a system of tubes is present serving to convey fluid to the most distant parts of the plant. This is the way plants drink water, and with it take in other materials, such as *nitrogen, hydrogen, sulphur, phosphorus* &c., in the form of *nitrates, sulphates, phosphates* &c., which are essential for the building up of living tissues of plants.

Leaves also considerably help the upward flow of the water. But to be able to properly understand what part leaves take in this difficult business of raising water to such a height, we must first learn something about the leaf, and we cannot do it better than by studying that of the *ashwathwa*. It has two parts; the larger part is thin, broad and flat, with a network of ribs and veins and terminated by a pointed end: the smaller part is the stock, which is round, smooth, somewhat long, and very slender causing the leaf to tremble at the slightest breeze. During the greater part of the year the leaf is green of varying shades. The surfaces of the leaves are not continuous but are iterrupted by great numbers of small openings which allow air to enter into and escape from a great system of interspaces which is present in all the tissues. Owing to this air is constantly enterning the plant by the leaves and water evaporating from them, unless when the atmosphere is salurated with vapour. The air contains carbonic acid which serves as food for the plant, because the living matter of the tissues has the power of decomposing it and retaining the carbon whilst the oxygen is set free; and the evaporation of the water plays the part of a suction pump and helps to draw the water upwards from the roots.

We have been led to make these remarks in order to show that the leaves take an important part in raising water in a tree. But we have yet to see how this is done. It is simple enough. The living contents of leaf are always thirsting for water; so that, as soon as any cell loses water by evaporation, it sucks it up from its neighbour, and this process of robbing the neighbouring cells of their aqueous contents goes on down to the very extremity of the root and the water is drawn up, as it were step by step, to the top.

If the under surface of an *ashwathwa's* leaf be examined with a powerful lens, or better still, under a microscope, a large number of pores will be found. They look like so many mouths, and to make this resemblance to a mouth complete, the two cells situated at the entrance to the pore look remarkably like a pair of lips. Each of these pores open into an air cavity mentioned above. It is through these pores that the surplus water of the leaf escapes into the air as vapour.

The openings in the surfaces of the leaves are so constructed as to regulate the escape of moisture. Nothing is purposeless in nature, much less these lips. They perform some very important function in the economy of our *ashwathwa's* life. They are the veritable turnkeys of the pores of the leaf, through which water escapes into the air as vapour. During the fierce heat of summer when everything is dry, and water becomes scarce, the large and wide spreading *ashwathwa* has to be very economical as to the way in which it uses the water. At such a time the lips render the tree useful services. They come together and thereby close the pores, and diminish evaporation. Again when the rain sets in, and the earth becomes moistened, and the roots suck up water, they stand aside to open the door of the pores as it were, and allow the evaporation to go on as before. The ancient Hindus were wise in laying down sacred injunctions to irrigate the more useful plants during the dry hot month of *baishakh* and thus supply them with moisture at a time when they need it most.

How the ashwathwa feeds and digests. We have already learnt that carbonic acid serves as food to the *ashwathwa*, and that the leaves obtain it from the floating air. We have also learnt that water is always present in the leaf, and for that matter, in every part of this giant tree ; and that there is some kind of green matter in the green cells of the leaf. We need not trouble ourselves as to what the name of this green matter is ; suffice it to know that it is essential to the life of the *ashwathwa*, and needless to say that it plays an important part in the production of those substances of which the tree is composed. This is what happens. There is the green stuff in the cells of the leaf on the one hand, and a supply of carbonic acid derived from the air, and of water derived from the soil on the other. Under the influence of sun-light living matter containing this green substance is able to effect certain chemical changes, in the course of which carbonic acid is decomposed, the oxygen being set free and the carbon entering into combination with the material present in the water derived from the roots &c form compounds such as starch, sugar &c. These are dissolved in the sap and in combination with nitrogen and other substances serve as the sources from which new protoplasm arises, and from which new cells and tissues are ultimately developed.

Sal, tal, tamal, AND piyal are trees of classic fame in India. Of these, the *tal* is the most familiar in Bengal and Southern India. As a child the *tál* tree was a great mystery to me, being at once puzzling and fascinating. I wondered how it was that the same tree which yielded delicious *talsans** and was the source of sweet cakes, also supplied those hateful palmyra leaves. To be candid, when the detestable village pedagogue set before me the impossible task of scribbling twenty of these leaves before breakfast, I secretly but sincerely wished the whole race of palm trees extinct. But my troubles came to an end when Janardan took me in hand, and taught me to take an interest in the "how" and "why" of things around me. In a few days I learnt many things, and forgot that reading or writing was a task; as for getting by heart multiplication tables, I found it the easiest thing in the world. The magic power of sympathy and judicious guidance brought about this change, not in me alone, but in all those who ever came under the influence of that honest soul—Janardan. The *tal* tree still continued to exercise my mind. It was a tree to be sure, thought I, but its bare gaunt stem and circular head of leaves were so unlike those of ordinary trees. Janardan had noticed that I looked curiously at the palmyra trees and brooded over something. He asked me one day if anything troubled me, and I told him what I thought about these trees. He did not laugh at me, or call me silly and sentimental. Oh, no, he was too good

---

* Immature seed of the fruit.

and sensible to make light of a boy's troubles. On the contrary, he remarked that that was the right way to look at things, and took occasion to explain to me that there were a great many other trees, such as the cocoanut, the date palm, and the common *supari* (areca nut) which had bare branchless stems, and in which the leaves, though differing in size and form, were arranged in a circular head.

This knowledge about the difference in the forms of stems was crude and elementary, but as a mental discipline it stood me in good stead afterwards. But that is another story. Despite good Janardan's explanation, the puzzle of the palm tree remained unsolved with me; as, however, there were other things to occupy my mind I did not trouble myself much about it until Svamiji let me into the secret of the matter. It came about in this way.

There were a number of very large and very lofty palmyra trees in Svamiji's grounds. Every year large numbers of weaver birds (*baya*) resorted to them for nesting, and from May to August a great many of those long, elegant, retort-shaped nests could be seen depending from every one of these trees. I was very fond of watching the tiny architects—the weavers, at their work, and often spent hours, sometimes alone, but more frequently in company with Janardan, near the palmyra grove. One of the nests was nearly finished; the hen bird had taken her seat inside; the cock bird was working outside, pushing the fibres here and there, and the hen bird arranging them according to her taste and judgment. All of a sudden the cockbird left the nest, disappeared for a few seconds, and then came back with something in its beak. It next busied itself for a minute or two, and then flew away as abruptly as it did before. While the bird kept on thus flying in and out, Janardan asked me if I knew what this restlessness of the bird meant. I did not know of course, and said so. He explained to me that every time the bird came in, it carried a very small lump of mud in its beak which it employed in plastering a portion of the nest. Opinions differed as to the object of this plastering. Some thought it was to stick fire-flies on to illumine the nest at night. Janardan did not believe the story, but advised me to observe it for myself and find out whether there was any truth in it; others said that they provided these lumps to sharpen their beak. A few were of opinion that the mud was employed to strengthen the nest and balance it properly. While I was deeply absorbed in listening to Janardan's stories about the nesting habits of these birds, and admiring their untaught skill, somebody gently tapped me on the shoulder, I turned round, and saw the venerable figure of Svamiji before me. "Well, my boy," said he, "has the palmyra tree told you why it chose to have such a bare and tall stem?" I felt ashamed and confused. I was no longer a boy, and a desire to appear what I was not, call it conceit or anything else, had already cast its darkening shadow upon my mind, so that I felt annoyed that Janardan had betrayed my secrets to Svamiji. But it was for one brief moment. One look from those kindly eyes inspired me with confidence, and assured me that I had nothing to fear, nothing to be ashamed of. "Question the trees," said he in an encouraging tone, "and they will tell you all about themselves." Here was an enigma: how was I to question trees? Was he joking? But before I could stammer out a word or two, he gave me a few grains each of paddy, barley, gram, *mung*, and some tamarind seeds, with instructions to sow them and water them, till the seedlings were about four inches long, adding that they would then speak to me. I sowed the seeds with care, and watered them with assiduity; they germinated and were about four inches long, yet they spoke not. Disappointed, I went to Svamiji. He did not, however, appear to pay much heed to what I said, but told me to prepare rough drawings of the seedlings with accurate descriptions of their parts. I may mention in passing that elementary drawing was then recently made compulsory in the Vernacular and university Entrance examination. I sketched and described more than once, and although Svamiji praised me for my willingness and perseverance, my performance evidently did not meet with his approval. Every time I showed it to him, he pointed out some error of observation here, suggested a little alteration there and talked to me in such an encouraging manner as to enkindle in me fresh enthusiam to try it again. "Well done, my friend!" exclaimed he in great delight, when I showed him for the fourth time the sketch and description, "that is the proper way to question the plants, let us see what answers they have made."

Seeds of gram, *mung*, and tamarind were, in growing, *split up into two halves* ; and the growing points had taken opposite directions of the same axis—*upward* and *downward*, the part which had taken the upward direction becoming the young stem supporting the two true leaves, the downward part which assumed a tapering form becoming the root. Seeds of paddy and barley were *not split up into two halves*, and there was no downward prolongation of the growing point, so that, instead of there being one primary root of a tapering form, there were many of them, all pushed out from the base of the stem. Svamiji explained to me that a large number of flower-bearing plants, among which were included the *ashwathwa*, the banyan, the mango, and a great many other common and uncommon trees, shrubs and herbs belonged to the former, or, *two-seed-leaf* kind ; on the other hand, a large number of plants including various species of palms belonged to the latter, or, *one-seed-leaf* kind.

I was now perhaps convinced, Svamiji remarked, that the *ashwathwa* and banyan differed somewhat in their fundamental plan of construction from the palmyra, the date, and the cocoanut palm. As a further confirmation of the fact of this difference, he made a small cut on the stem of a banyan causing it to bleed ; but although the same operation was performed on the stem of a palmyra, no trace of moisture was found. The reason of this was, he explained, that whereas the outer fibres of the banyan retained their vitality and power of growth, those of the palmyra had ceased to grow with the fall of the leaf with which they were connected. The leaves of *ashwathwa* and banyan &c. fall off, and new ones take their place; the palmyra and other trees belonging to the same class, also shed their leaves, but here no new leaves take the place of the old ones. The parts with which the old leaves were connected become dead, and the nutriment which sustained the old leaves is carried further upwards. Hence the stem instead of growing thick and branching, becomes tall, thin and gaunt.

The *Piyal* and *Tamal*, beloved of gods and poets, are not quite so familiar as the palmyra, or, the *ashwathwa*; yet, they are not at all uncommon trees in Bengal and other parts of India. A native of the mountainous forests of the coast of India, the *piyál* is also found pretty well distributed all over the country, especially in Mathura and Brindaban. It is a superb tree, with a straight, thick and lofty trunk supporting numerous wide-spreading branches, with smooth oval oblong leaves, and bearing during the spring a profusion of whitish green flowers. *Tamal* is found in most parts of India. It is a tree of a middling size, with lengthened heart-shaped leaves which are covered with a kind of downy stuff. It is first cousin to one of our well known and very useful trees, the *gab*, and hence popularly called *bangāb*.

For its elegant and graceful form and its fragrant flowers, the *bakul* is a universal favorite with the Hindus. It adorns the rich man's garden, shelters the poor man's cottage, and forms an appropriate adjunct to holy shrines and *mathas*. There is scarcely a town or village of any pretension in India which has not a few fine trees as places of public resort, and, more often than not, one of them is a *bakul*. How well do I remember those delightful summer mornings, when half awake, half asleep, I used to walk with nimble steps towards the *bakul* tree of our village to gather flowers which lay thickly scattered on the ground. Great would be my sorrow if, by chance or ill luck, I was late any morning, and found the merry companions of my boyhood already in the field. But we were not the only competitors for the spoil; with the sun-shine came bees aud butterflies to taste the sweet nectar which lay hidden in the tiny cup of the flower. But it is impossible to look back upon those days without a shade of grief and shame at the thought that in spite of my love and fondness for the *bakul* flower from childhood onwards I never cared to study it carefully, and examine and admire its beautiful structure, until the venerable Svamiji opened my eyes to look at it in a new light. It was only then that I realized that the sweet fragrant little flower is a marvel of skill and workmanship, its every part being admirably adapted for the particular work it is called upon to perform. Outside, it is protected by a small green cup or case with a double series of tiny leaves; next to them is the second floral envelope consisting of a small short tube with segmented border, the segments assuming the form of very small leaves. At the mouth of the tube are placed small ragged and hairy conical bodies called *nectaries*, in which the honey is stored; inserted alternately with these nectaries or store rooms for honey are eight short hairy filaments bearing at their top linear sharp-pointed bodies full of a kind of powdery substance. On seeing with a lens, a germ will be found; it is divided into eight cells with an ovule in each.

Kadamba is another tree of romantic association to a Hindu. Its very name takes him across ages to those happy and joyous days when Krishna flourished in Brindaban. It was under its delightful shade redolent of sweet perfume that the pastoral god danced to the soul-stirring music of his lute, breathing such divine strains as to cause man and beast to stand hushed and wondering. But to appreciate the glory and beauty of a *kadamba*, we must observe it in midsummer when the tree blossoms, and bears a profusion of those highly ornamental and delicately fragrant flowers so unique in the vegetable kingdom. The next time we come across a *kadamba* in full blossom, we must not pass it by without making a closer acquaintance with its flowers, which form a large perfectly globular head of an orange colour with large white club-shaped stigmas standing out.

Who ever thought that the common *gandharaj* (Gardenia florida) with its richly fragrant flowers has any kinship with the stately and graceful kadamba? Yet, it is so, as an examination of their leaves and flowers will prove. Though one of the favorite shrubs in almost every Indian garden, the original home of the *gandharaj*, according to a learned and distinguished botanist, is China, whence it must have been introduced into India at a very remote age. But it is not the *gandharaj* alone that claims kinship with the *kadamba*. There are numerous other plants which belong to the same order, some of them of great economic value, as for instance, the cinchona, from the bark of which the far-famed quinine is obtained. Invaluable as an article of sick diet, the *gandhabhadali* (Oldenlandia alata) must be familiar to many. An examination of its small white flowers which appear during the rainy season will at once reveal the fact that it is closely allied to the cinchona and the *kadamba*. Among other plants belonging to the same order may be named the *manjista* or Indian madder, of which the roots, stem, and larger branches are extensively used for preparing a kind of red dye.

There is yet another tree which it is impossible to pass by without admiring. It is the *asoka*, or the (sorrowless) of the ancient Indians. The name is very appropriate indeed, as it is impossible to look at an *asoka* in blossom with its clusters of beautiful flowers of varying shades of yellow and orange without being filled with the gladsomeness they shed around them.

There is a fine avenue of *piyal, tamal, bakul* and

kadamba in Svamiji's park. Overhead their wide-spreading branches meet and interlace one another, forming a shady canopy so dense that, even during the fierce heat of an Indian summer, the scorching rays of the sun reach the ground subdued and broken into a thousand fragments, to give light and life to the innumerable species of ferns, aroids and other undergrowths with which the grounds abound. Here, during the livelong summer days, cuckoos answer cuckoos, and, fairy-like, the paradise flycatcher flits about from bough to bough or dances to the music of the *doyal* and the bulbul. Leaving their mountain home behind, flocks of gay warblers, with the merry companions of their migration, the bee-eaters and fly-catchers, winter here, and enliven the aspect of the park by their sprightly cheerful ways. Sitting in the midst of thick foliage, and screening itself from the profane gaze of man, the tiny *chatak* (white-winged green bulbul) utters its long-drawn mournful note, supplicating, it is said, for a drop of water. Be that as it may, there is something mysterious in the sound which carries with it a sense of gloom rather than cheerfulness to the mind, especially in the still hours of a summer midday.

But the most striking feature of the grove on a bright windless summer day is the abundance of butterflies. Their conspicuous beauty, elegant forms, and rich and varied colours can hardly fail to strike even a casual observer. Maddened with the pleasure of mere existence, a vast multitude of them flutter and dance in the air, performing fantastic evolutions; others hover about trees and bushes, wooing buds, kissing flowers, and sipping honey.

> Go, child of pleasure, range the fields,
> Taste all the joys that spring can give.
> Partake what bounteous summer yields.
> And live whilst yet 'tis thine to live.
> Go sip the rose's fragrant dew,
> The lily's honey'd cup explore,
> From flower to flower the search renew,
> And rifle all the [*asoka's*] store.

VIII.

THE HOME FARM.

We have been spending, of late, a good deal of our time in the plantation, and, as good luck would have it, in the company of a most interesting and worthy man. By profession he is a sailor, but, having become old, and a family having grown round him, he has given up "sailoring," and settled down to the peaceful occupation of a cultivator and farmer. Rājarām, for such is the name of the man, is the only son of a poor but respectable fisherman who plied his vocation on the river Hooghly near Calcutta. While yet a lad, he took to accompanying his father in his fishing excursions, and spent more time on the waters than at home. Being intelligent and willing, he soon learnt to manage nets and fishing tackles with such deftness as to please his father and surprise his kinsfolk. There was, however nothing singular in all this. As a race, fishermen are bold and hardy all the world over, and a Bengal fisherman is no exception to the rule. Growing from a "sprat to a whale" as the saying goes, upon the waves, he loses all fear of the water, and becomes resourceful from an early age. As contemporaries of our hero, there were at least a score of juvenile fishermen employed in similar pursuits, and some of them were capital hands at the business. They paddled the boats, cast and set nets, gathered in the day's haul, smoked *hooka*, ate their coarse but substantial meal, slept under the broad canopy of heaven, and rose refreshed to begin their daily work again as did their fathers and forefathers before them, without a thought beyond the narrow sphere of their simple life. But not so Rājārām. He would rise superior to the traditions of his profession, strike out a new path, and enter, as it were, upon a new phase of existence. His natural surroundings greatly favoured this bent of his mind. Brought up in the midst of ships and sailors in the Calcutta port, he had early conceived a strange passion for a seafaring life, which took a stronger hold of him as he emerged from boyhood to youth. Often while steering his father's frail boat, or paying out the interminable lengths of a *berh jāl*, he would watch with closest attention the officers and crew of the sea-going vessels engaged in their respective duties, and would secretly wish to be one of them. The clock-work regularity with which everything was performed on board a steamer, the far-reaching whistle of the boatswain, the musical click of the capstan working, or the mellow tinkling of the bells, fascinated Rājārām. But, like the good fellow that he was, he bided his time, and waited for an opportunity, which was not long in coming.

One forenoon, during a certain passenger season, (to specify particulars is unnecessary,) one of the popular steamers of the splendid fleet of the P and O line was full of passengers and their friends and relations. When all were busy talking and leave-taking, a lovely little girl of seven summers with a few friends, boys and girls about her own age, was amusing herself by throwing bits of bread and biscuits into the water to tempt the kites and crows. In doing so, she on one occasion leaned forward a little too much, lost her balance, and fell overboard. Great was the consternation among the passengers and crew, and every body rushed forward where others had preceded them. Unspeakable was their joy to find the girl safe, but dripping like a mermaid, on a small fishing boat. The credit of the gallant rescue belonged to Rājārām. His boat was just gliding past the steamer, and while watching, as was his wont, what was going on on board the vessel, he had observed the girl falling, and, without a moment's hesitation or loss of time, had plunged into the water after her. That the plucky young fisherman should not wait for *buksheesh* was, of course, a matter of surprise to every body. But as it was just time to weigh anchor, nobody could spare a thought about it in the hurry and confusion of the moment.

At about 11 o'clock next morning Rājārām was ushered into the presence of the representative of the Peninsular and Oriental Company in Calcutta, a high souled gentleman who had sent messengers in all directions to find out the whereabouts of the young fisherman who had saved the life of his little niece, with instructions to bring him to his office when found. "Brave young fellow" said he addressing Rājārām, and offering him a purse containing two hundred rupees in silver, "here is your reward for your humane deed."

"My good sir," answered Rajaram humbly but firmly, "I deserve, I am sure, no reward for what I did. Is it possible for a human being to witness an innocent child drowning, and pass by without giving help? No sir, I do not think I did anything extra-meritorious." The gentleman to whom the above little speech was addressed was greatly impressed by its frankness; besides, there was something so out of the common in Rājārām's look and manners that, instead of being annoyed at his refusal to accept the proffered reward, he was much pleased with his honest and manly answer. Shrewd, sensible, and kindly in disposition, he soon managed to draw Rājārām into conversation with him, and in half an hour made him disclose what his aspirations and yearnings were. It was an eventful day for our hero: the dream of his life was about to be realized! He was taken on as an extra hand on board the S. S. "Nepal," proved himself an able and willing helper, and soon became an able-bodied seaman. I must not omit to mention a little incident which occurred just at the dawn of his new life, and which made a lasting impression upon Rājārām's mind. Before a large assembly of ladies and gentlemen gathered together on board the "Nepal" bound homeward on her fifty seventh voyage,—on that memorable day of his life, when, amidst doubts and fears of his friends and relatives, he was about to launch into an uncertain career Rājārām was presented by a distinguished lady a bronze medal bearing the following inscription :—

"Here and there a cotter's babe is royal born by right divine."

He has sailed many seas, and seen many lands, and is a mine of useful information. He is altogether an original character, and, like many other original characters, is at times eccentric and irritable. While ever ready to do a good turn to the deserving and needy, he has no patience with impostors and evil-doers. Unscrupulous petty landholders given to tyrannizing over their poor tenants,—and unfortunately there are many such,—he detests. I have seen him greatly agitated with rage and scorn at the mention of the name of a certain individnal, who, by birth a Brahman and by profession a clerk, is at heart a ruffian and a knave, and who lately cheated an old widow of her homestead lands. Rājāram would think nothing of travelling six miles in rain and storm to succour a poor peasant who may have a calf or a cow ailing. This puts me in mind of his treatment of his cattle and farmstock. For every one of them he has a name, and they share his affection and attention with his children. He is never weary of impressing upon the minds of his children the necessity of kindness towards our domestic animals, and rightly holds that those who are unable or unwilling to treat them well ought not to possess cattle. " It is sinful," says he, "to keep animals for our profit or pleasure and yet not to study their comfort and well-being." One can hardly imagine what an amount of time and trouble he bestows upon his domestic pets. Up at five, in the early morning he regularly employs himself, with the help of his equally willing and industrious sons, each of whom has his allotted share of duty to perform in the service of his farm stock. Stalls are cleansed, feeding and drinking vessels scoured and washed, cattle groomed, and food suitable to the taste and requirement of each prepared; and everything is performed in a most cheerful spirit and in a business-like way without fuss or talk. Grooming a cow or bull sounds unusual, as we are not accustomed to see it practised; but Rājārām assured me that it is very good for the cattle, as it keeps their skin and coat clean and healthy, and improves their appearance. I am perfectly convinced of this; a finer herd of cattle than those owned by Rājārām perhaps does not exist anywhere in Bengal. They are quite different from the small-boned miserable creatures we see everywhere about us.

As a suitable tribute to his great kindness towards them, the cows make him a good return in large quantities of rich milk, and the bullocks in labour. Rājārām has been the indirect means of creating a spirit of healthy emulation among the agricultural community of Basantapore and its neighbourhood, and many have begun to appreciate his way of treating cattle and farmstock, and imitate him in setting apart pasture lands and raising fodder plants for them. Even the apathetic and cynical have now been convinced of the importance and benefit of cleanliness, liberal food, and good treatment in general in the management of dairy and farmstock. It has been brought about thus. A wide-spread cattle disease broke out last year in Basantapore and the neigbouring villages. Hundred upon hundreds of cattle died every week. The people were powerless to arrest the progress of the disease, and suffered stolidly what they could not remedy. There was, however, one man who did not lose a single head of cattle, and that man was Rājārām. From the very commencement of the outbreak, he adopted energetic and efficient means to save his worldly wealth. He built some temporary huts in an isolated tract of land, and removed his stock there; kept up an almost continuous fumigation for weeks together, and cut off all communication with his house which stood within the affected area. In-order to enable his animals if attacked, the better to withstand the depressing influence of the malady, he added to their liberal diet such stuffs as experience and the knowledge of cattle had suggested to him. The result justified the means: except a few cases of *diarrhœa*, nothing serious happened to his stock. This escape, which seemed miraculous to the ignorant and superstitous, was naturally talked about a great deal; and while lazy and foolish people ascribed it to his incantations and supernatural powers, and good luck in general, the intelligent portion of the community rightly set it down to his superior and sympathetic treatment of his farm stock. Most beneficial has been the result of Rājārām's good example; a decided change for the better is noticeable in Basantapore and its neighbourhood in the treatment of cattle.

Rājārām is never weary of expatiating upon the

wisdom of the ancient Hindus in laying down sacred injunctions designed for the preservation and welfare of cattle, or upon the degeneracy of their descendants in neglecting to observe them. He is right in attributing the deterioration of cattle to the great apathy of the educated and well-to-do Indians who, whatever their other merits may be, take not the least interest in matters which concern the requirements of human existence. There is no want of deep-rooted faith or sentiment in regard to the treatment of cattle; but, what Rājārām deplores is the absence of their proper application under intelligent guidance. While in England, he took the opportunity of visiting various cattle-breeding establishments, some of them owned by distinguished noblemen, who devote not only large sums of money, but much time and energy, to the improvement of the various breeds of English cattle. We wish there were a few such noblemen in India.

THE FARMSTOCK.

Rājārām has travelled all over India in search of good cattle, and, as a reward of his trouble, has the satisfaction of being the owner of an exceptionally fine herd of bovines. To begin with, he has three or four strong healthy Indian oxen, the so-called Zebus or the Brahmani bulls of the Europeans in India—shapely animals with well-developed, firm humps, and large dewlap. They render important service as sires. His sleek and fat team of Amritamahal bullocks will gladden the heart of even the most fastidious judge in any cattle show. The Amritamahal breed is no doubt one of the finest in India, and its fame dates from the time of the Hindu princes who ruled Mysore before Hyder Ali of historic renown. Hyder Ali himself, and his son, Tippoo Sultan, did much towards the improvement of the breed. As draught bullocks they have no equal. Besides these, Rajārām has imported cows from Hissar known as Hansi breed: they are excellent milkers. He has Santhal cattle from the Burrakur river; the Purbi cattle from Allahabad; the Gorannea of Bundelkund, the Bagondhá of Oudh, and a breed of small cattle from Jessore. Though not very active, the Nellore bullocks, with short sharp horns, pendulous ears and massive frame, are imposing looking animals useful for slow work. The cows yield good milk rich in cream. They are very good-tempered and tractable beasts. The Krishna and the Guzrati bullocks are also very fine-looking noble animals, but slow. As milchcows there is perhaps no breed that can equal those from Junagar in Kattiwar. It is not alone for his own profit or pleasure that he has got together this fine herd of Indian cattle; his great object is to demonstrate the fact that for new blood to improve the stock we need not go outside India

IX.
INDIAN SNAKES.

India abounds in snakes. Learned European scholars, who have paid much attention to Indian ophiology, have determined that there are about two hundred and sixty-four known species of snakes in this country. Some of them, such as the cobra, the krait, and the daboia &c. bear a deadly venom; the rest, or at least the great majority of them, are harmless. But, whether harmless or venomous, snakes as a class have, from time immemorial, been looked upon with a prejudiced eye and therefore abhorred and shunned. Yet there is a certain fascination in snakes, and, in spite of our dread and hatred of them, we like to look at their lithe graceful forms and sinuous movements. Be that as it may, our object is to attain some knowledge of snakes; we must therefore, contemplate them as members of the great animal kingdom, and endeavour to study their wonderful organisation and habits.

To a student beginning the study of snakes, the first question that suggests itself is, *what relation does a snake bear to other animals?* We must allow that the question is a pertinent one. But before we proceed to answer it as best we can, it is as well to warn all serious students of Natural History that "no knowledge can be attained without study," and least of all a knowledge of Natural History, to acquire ever so little of which one must be prepared to take infinite trouble. From this point of view, we should advise those who are interested in the study of snakes to supplement their knowledge of books with that derived from a practical acquaintance with the structure and habits of these interesting reptiles. Unless this is done, one's ideas about them will always remain hazy and imperfect. There are several ways in which this can be accomplished. It may be mentioned in passing that there are European youths in India, who, in the midst of their engrossing business avocation, have by dint of perseverance and industry, succeeded in making interesting collections of natural history objects, and acquired a fair knowledge of the zoology of our country.

Snakes considered in relation to other animals. The animal kingdom is divided into two main sub-divisions, or sub-kingdoms as they are called in zoological language,— *Vertebrata*, or animals with a skeleton, and *Invertebrata*, or animals without a skeleton. This, it must be understood, is a popular classification. The sub-kingdom Vertebrata is again sub-divided into *Mammalia*, or sucking animals; *Aves* or birds; *Reptilia*, or reptiles; *Batrachians*, or frogs, toads, and newts; and *Pisces* or fishes. The class *Reptilia* is further sub-divided into *Crocodiles, Tortoises, Lizards,* and *Snakes.* It will appear from the above scheme of classification that however repugnant the idea may be to us, there is no help but to accept the fact that we are remotely related to one another in that both man and snake belong to the same sub-kingdom, *Vertebrata*, with innumerable intermediate gradations of structure and organization separating them.

Characteristics of the snake. A snake may be briefly described as an apodal or limbless modification of the vertebrate animal with an elongate body. Of course, the most obvious and striking characteristic of a snake is to be found in the nature of its organs of locomotion. Our idea of progression in vertebrate animals is associated with limbs, or some modification of them; in the snake, however, we come to find a departure from this established plan of creation. But to be able to properly understand how, in the absence of limbs, the snake can move about as freely as any other animal, we must have some knowledge of its structure. An ardent student of natural history will try to acquire this knowledge by dissecting a snake and making himself familiar with its anatomy, or, as the next best substitute of it, by visiting the "Reptile gallery" of some museum.

How does a snake move in the absence of limbs? A glance at the skeleton of a snake will show that the ribs take the place of limbs in the act of progression in the snake. These ribs are very numerous, numbering more than three hundred pairs in some species, and are perfectly free at their extremities. Free, because, unlike a great many other animals, there is no sternum or breast bone in the snake. Being free and unattached, they are also very movable. The free extremities of these numerous and extremely movable ribs are attached to the abdominal shields, or breast scales by elastic cartilaginous filaments. Another thing which renders the movement of the ribs easy is the great moblity of the vertebral column.

Body covered with scales. The body of the snake is covered with scales; but a little observation will show that, all of them are not of the same size and form. The scales of the head assume the form of plates or shields, and do not overlap each other like those of the other parts of the body; those of the under surface of the body are broad and long, extending from side to side. The number and character of scales of the different regions of a snake's body are important elements in the determination of its specific identity.

Eye, tongue and teeth. Every body must be familiar with the forked tongue of the snake which is constantly darted in and out with a rapid quivering motion. It is protruded through a slit in the lower lip. A snake's tongue has not much to do with taste, which, as far as we know, is the proper function of this organ in other animals. What is the good then, it may be asked, of its having a tongue at all ? A great deal. As surely as a blind man gropes his way with the help of his stick, so surely does a snake feel its, with the aid of its rapidly quivering and sensitive tongue, in dark holes, in the midst of tangled weeds, and in all sorts of *debris*, which form the natural surroundings of a snake's life. The tongue of the snake when retracted within the mouth is lodged in a sheath. The chief peculiarity of the eye of a snake consists in the absence of lids properly so called; a layer of transparent epidermis extends over the eyeball and takes the place of lids: it is periodically cast off with the rest of the epidermal covering. Having no eyelids, snakes cannot wink, so that "the stony winkless stare of the snake," is not altogether a myth.

Mouth and teeth of harmless snakes. We now come to the consideration of the mouth and teeth of the snake, and, for the sake of convenience, take the harmless snake first. A snake seldom opens its mouth except for the purpose of seizing its prey, or in defence; or sometimes when yawning after food or drink; or again, as it has been observed to do in captivity, when the mouth is sore. If the upper and the lower jaws are separated, it will be found that one surface fits exactly into the other in every detail. It is thus brought about. There are four rows of teeth in the roof of the mouth which divide the palate into three elongated depressions. These three depressions receive the three corresponding elevations in the lower jaw. The elevations in the lower jaw are caused by the two rows of teeth on two sides, and the wind-pipe in the middle. Almost every harmless Indian snake has six rows of teeth,—four in the upper, and two in the lower jaw. In some snakes, all the teeth are equal or nearly so; in others, they are irregular. They are all directed backward, thus offering a formidable obstacle to any resisttance on the part of the prey when once seized. The mouth of the snake is, in fact, an efficient prehensile apparatus.

Mouth and teeth of a venomous snake. Having glanced at the characters and arrangement of teeth in harmless snakes, we proceed to the examination of those in the venomous snakes of India, taking the cobra, the deadliest of poisonous reptiles, as the subject of our study. On examining the mouth of a cobra (*keutiah* or *gakhura*), we come across the following peculiarities :—two slightly curved fangs on each side of the upper jaw; slight depressions in the lower lip for the reception of the fangs when the mouth is shut; absence of any teeth, except one or two rudimentry ones, behind the fangs. Also the fangs are not bare, but enveloped in a fold of mucous membrane; they occupy the same places as the canine or cutting teeth of dogs, cats, or other carnivorous animals. Unlike those in a non-venomous snake, there are only four rows of teeth in the cobra, two in the upper, and two in the lower jaw. If, now, the skin of the cheek be dissected away, from the nostril in front to the angle of the mouth behind, a comparatively large flask-shaped body, somewhat resembling the inner core of an onion, will be exposed; it is the poison gland of the snake. A small duct conveys the death-dealing fluid to an orifice at the base of the grooved fang, down which it flows into the wound which the animal inflicts. It may be mentioned in passing that expert and wily snake-men, the so-called *snake-charmers*, with a view to destroy the communication between the gland and the fang, cut through the duct instead of removing the fang, as is more commonly done. The dodge is very clever indeed, and serves its purpose admirably—that of deluding the ignorant and credulous public. It has been indicated above that there are one or two rudimentary teeth behind the fangs. These serve as a sort of reserve, so that if, by accident, or from any other cause the fang in use is broken, the small rudimentary one moves forward and takes its place and serves its purpose.

A great many other venomous snakes inhabiting India and its dependencies possess the same kind of poison apparatus as the Cobra. The *Sankhachur*, (King cobra, or the Hamadryad of Europeans in India); the *Kareta* (Krait); the *Raj sap*, (Banded krait); and the sea snakes belong to this group.

But it is in the Viper kind that the poison apparatus has attained its perfect development. The *Chandrabora*, (Russell's viper) is the most typical representative of this group in India. Its fangs are very long, and lie prostrate along the jaw, but are capable of full erection. To raise these long fangs, the bones of the jaw are so modified as to act like hammer-heads with claws representing fangs downwards. Normally the fangs are covered with a thin sheath of membrane.

VERY SMALL SNAKES.

The Typhlops, called *sapala* in Hindustani, and *puyasāp* in Bengali, are the smallest of all the known species of Indian snakes. The smallest of the group (Typhlops beddomii) inhabiting mostly the plains and hills of southern India scarcely exceed five inches in length. All of them prefer moist shady spots, and are found in decayed wood, and tumbled-down buildings.

A typhlops looks uncommonly like an earthworm, but a moment's careful observation will show that whereas the body of an earthworm is segmented or composed of rings, that of a typhlops is covered with minute scales. A few years ago, some of these tiny ophidians found their way into the system of pipes laid in the city of Calcutta for the distribution of pure drinking water; there, they bred and multiplied and caused quite a scare among the inhabitants, as they kept on dropping here and there from water taps.

SOME LARGE SNAKES.

The Python is the largest of all Indian snakes. It is sometimes, but erroneously, called *Boa constrictor*, which is an American snake. Two species of pythons are found in India and its dependencies, (1) the Rock snake, (*Moial or borāchite*) which grows to about 20 feet long, and inhabits various parts of India, especially the Sunderbuns, the forest clad slopes of the hills of Assam, the Terais &c, (2) the Reticulated Python which is a much larger snake, growing to about 30 feet in length, and found in the forests of Burmah, the Nicobars, the Malay Peninsula and the Archipelago.

Fanciful stories are current about the length and thickness of these snakes. Their power of seizing and devouring large animals has also been much exaggerated. A large python can seize and devour deer, goats, dogs, and animals of that kind, but not a buffalo as shown in picture books. It lies in wait quietly and patiently for hours together, generally with the tail coiled round something. On the approach of the luckless prey, it suddenly darts forward and secures it in its fold in the twinkling of an eye. So quickly is the whole operation performed that the eye can hardly follow it. If the animal seized be a strong and active creature, the snake presses it within its fold, and the muscular power, which it now exerts, is quite apparent from the wave-like motion of the upper part of its body. The snake begins to devour its prey head foremost. It is quite a mistake to suppose that the python, or, for that matter, any other snake, licks its prey before devouring it so as to smear its body with saliva.

THE PYTHON IN CAPTIVITY.

In captivity, a large python generally lies coiled up in a heap. It seldom moves about, except from bed to bath or bath to bed, or when hunger compels it to look about for food. During the hot weather it remains a great deal under water, spending a week or ten days in the bath without stirring. If unaccustomed to the presence of man, as all newly-captured pythons are, it resents the approach of visitors near the cage, darts furiously and repeatedly at the glass or wire, causing serious injury to itself. For a python sixteen to twenty feet long, a duck or a fowl once a week during the feeding season is enough. The feeding season lasts from March to November, a little early or late according to individual peculiarity. A python of this size can easily devour a goat, pig or deer, and does so in its wild state; but, in captivity, it is obliged to be content with easily obtainable and less costly fare, such as duck, fowl, rabbit &c., or, on rare occasions, kid. Smaller pythons feed on rats, chickens, guineapigs &c. Pythons in captivity have been known to become tired of the sameness of food. One accustomed to be fed on duck may refuse to go for it, but would readily take a fowl instead. During the cold weather and the early part of spring, pythons, like all other snakes, hybernate, *i. e.*, remain inactive without food or drink in a state of partially suspended animation. When deprived of its liberty, a python does not unfrequently remain in a state of sulkiness for a long time, refusing all offer of food. A python seventeen feet long has been known to go without food for a period of eighteen months, none the worse for this prolonged fasting.

Wise provision of Nature. A python feeding is a gruesome spectacle, and children had better avoid looking at it. Yet it is not without its lesson. The act of swallowing costs the snake much time and trouble especially if the animal to be devoured is large as compared with the capacity of its jaws; the mouth and throat become enormously distended and distorted out of shape. The question naturally suggests itself: how is the snake able to carry on respiration when the mouth is so full and the throat almost choking? The beneficent Creator has solved the problem in a most ingenious and, efficient manner. During the progress of swallowing, the windpipe of the snake protrudes from the mouth, enabling the python to carry on respiration as usual. What is true of the python is true of all snakes.

FALSE CHARGE AGAINST SNAKES.

Popular prejudice and ignorance give a bad repute to some of the common harmless snakes, such as the *Laodoga* (Dryophis mycterizans) and *Beata chiti* (Lycodon aulicus). Even a professional snake-catcher, unless he is an expert, hesitates to approach a *Bungraj* (Dipsas trigonata); its repulsive viperine aspect terrifies him. The origin of the false charge may be traced to the fact that these snakes possess long fang-like teeth which are very useful to them in holdinig tough-skinned prey, such as lizards, toads &c.

THE COBRA.

Of all the snakes now inhabiting the earth, the cobra is undoubtedly the best known. It has acquired a world wide reputation as the deadliest of all the venomous snakes. The cobra has a very wide range of distribution, being found all over India, Burmah, Ceylon, the Andaman Islands, southern China, Indo China, the Malayan Peninsula and the Archipelago. In the Himalayas, it ascends to an altitude of nearly 8000 feet. To the west, its range extends to Afganistan, North Eastern Persia, Southern Turkestan, as far as the Eastern coasts of the Caspian Sea. Within India, certain tracts are more favourable to its multiplication than others, although there is not a single village where, at one time or another, a cobra is not found. In Bengal, the Sunderbuns are the congenial home of the cobra, especially of the *keutia* variety. Parts of Durbhangah and Rungpore are much infested with these dreaded reptiles. Though popularly believed to be two different kinds of cobras, the *Gakhura and the keutia are mere varieties of the same species.* Gakhuras show a preference for high and dry ground, and generally keep to the neighbourhood of villages and homesteads. Sometimes they are found inside houses and are looked upon as custodians of concealed treasures. The *gakhura* is the favourite snake of the snake-charmers, as it is generally good-tempered, and its movements are deliberate, graceful and more amenable to control than that of the *keutia*. The distinctive mark of a *gakhura* is its double ocellus on the hood. *Keutias* are numerically more abundant than *gakhuras*. There is an immense variety of them, and they prefer low swampy and unfrequented places. A cobra of the *keutia* variety, especially if it is young, is more active and restless than a *gakhura*. During the rainy season, when large tracts of the country become flooded, *keutias* are reduced to great straits. At such times, they take refuge upon trees in large numbers and live, for the time being, in perfect amity with the other occupants of the trees.

THE COBRA IN CAPTIVITY.

In captivity, the cobra, like any other snake, seeks retirement. A young cobra is less retiring than an old one, and even likes to display its fully expanded hood. A cobra, coiled up in a branch and swaying its expanded hood, is a beautiful sight indeed. Some new arrivals have been observed to remain like this for a whole week. When a number of them live together, they are sometimes given to fighting among themselves, but seldom with any serious result. Even a careful observer will find it difficult to ascertain what provokes the fight. An experienced snake-man thinks nothing of going inside a cage full of cobras with fangs, after they have got accustomed to his presence for a few days. The only precaution necessary is that he must carefully watch their movements the whole time. Cobras feed on rats, frogs, toads, small birds &c. They do not, like pythons, coil round their prey. New arrivals sometimes remain in sulk for days together; some even starve themselves to death. During the winter, they generally lie inactive under the blanket or straw, occasionally one venturing out to seek warmth in the cheerful rays of the sun, if there be any. In summer, they move about to refresh themselves with the cool breeze.

THE SANKHACHUR

is another species of cobra, which is as remarkable for its size as for its curious habit of eating other snakes. It is the King cobra, or Hamadryad of Europeans in India, and is the largest of all Indian venomous snakes, attaining to a length of twelve to fifteen feet. It is a very powerful snake, and rather aggressive in nature, but, fortunately, it inhabits unfrequented forest regions, being seldom or never found near villages or towns. In spite of its strength and aggressiveness, it very soon gets used to captivity, and its captor comes to exercise wonderful control over its movements, having of course, previously taken the precaution of breaking its' poison fangs. Instances are known of full-fanged *Sankhachurs* performing movements at the command of their keepers.

The *kareta* (krait), the *Rajsap* (Banded krait), the *Chandrabora* or *Bejagur* (Daboia or Russell's viper) are other common venomous snakes of India. Of these, the krait is rare in Lower Bengal. It is common in the Santhal Pergunnahs, Behar, and the North-West Provinces, and, more or less, in other parts of India. A large percentage of deaths from snake-bite in Behar and the N. W. P. are attributed to this snake. Those generally seen are from three to four feet in length, although specimens obtained from west Midnapore have been found to attain a length varying from five feet to five feet six inches. In its habits, the krait resembles the cobra. Russell's viper is a most repulsive-looking snake. Its fangs are much larger than those of any other venomous snakes of India. It is very sluggish, and seldom moves away at the approach of man or beast. Snake-men never keep Russell's viper for exhibition, as its movements being uncertain, no control can be exercised over them.

Closely allied to the *chandrabora*, but smaller in size and less deadly in nature, is the carpet viper (*Echis carinata*), called *kuppur* or *hafai* in the Punjab, and *Phoorsa* in south west India. This aggressive little snake is common in the Punjab, Sind, Cutch, Rajputana, the North West, the Central Provinces, and Bambay, but unknown in Lower Bengal. By rubbing the fold of the sides of its body, it makes a curious prolonged hissing sound..

There are several other species of venomous snakes in India which belong to the same family as the rattle snake of America. The bite of some of them is said to be exceptionally fatal to man, but this requires confirmation. As most of these snakes inhabit hills and forests, and seldom appear near human habitation, their power of doing mischief is very limited. They are so unfamiliar that except one or two species, especially the "karawala" found in Ceylon and in the Western Ghats, their identity is unknown to most of the Indians.

THE SEA SNAKE.

Various kinds of sea snakes infest the coasts of India and its dependencies. They are all venomous and are easily recognised by the peculiarity of their tail which is flattened vertically. Their poison apparatus resemble that of the cobra, but poison fangs are smaller. With the exception of one or two, none of them ever leave the water. On rare occasions they come up the river Hooghly as far as Calcutta.

GENERAL REMAKS.

Unless compelled by circumstances, a snake, as a rule, seldom comes out during the day, but remains in a state of perfect repose in some unfrequented lonely spot; tussocks of grass, dry leaves and twigs, a fissure or a hole in the ground, or some such convenient hiding-place giving it shelter. During the summer, it creeps out of its concealment in the forenoon and goes for a drink if water be near; but it seldom ventures into the open, unless satisfied that the coast is clear. While drinking, it keeps its head under water. Like many other animals that live by preying upon others, the snake comes out at night, and, having satisfied hunger, retires to digest its food, to a cool breezy place in summer, and to a cozy retreat in wet weather. It must not be understood, however, that the snake never feeds during the day. For, who has not seen the common *hele sap* (grass snake) pursuing small toads and frogs in broad day-light? The snake seldom comes out during the winter. The most enjoyable time for a snake is the early part of the rainy season; it is very lively at that time. The habits of different species vary much as regards their choice of habitations. Amphibious snakes must be sought for in the neighbourhood of water, especially near old tanks and dirty ponds. The *keutia* and the *Chandrabora* affect low-lying paddy and grass fields; and the *Gakhura* dry places in the neighbourhood of human habitations, finding ready-made and convenient shelter in the old galleries formed by white ants and rats. When a snake takes up its quarters at a certain spot, it is difficult to eject the unwelcome intruder without killing it. The *laodaga* (green whip snake), the *betacher* (a species of Dryophis), the *bungraj* (a species of Dipsas) are very arboreal in habits. In fact, every terrestrial species, except the lowest types, is more or less arboreal. The snake casts its skin once every two or three months; the old skin becomes dark as the new one forms underneath, and the reptile remains in a state of dullness without food or drink during the period of moult. Its passage through grass, stones, or any resisting object, however slight, helps the snake to get rid of its skin. The first gap appears near the mouth, and, by gradually pushing through the resistance, it manages to disengage itself from its old coat, which, except the tail end, comes off reversed.

Much has been written and said about snakes fascinating small animals. But all great ophiologists, who have paid any attention to the subject, agree in pronouncing it to be a myth. What actually takes place may be best expressed in the word of a disinguished observer :—"Whether ground or tree snakes, they remain patiently in the same attitude until their prey approach ; then, gently gliding over the short distance which intervenes, they pounce on the unsuspecting victim. The approach is so imperceptible that doubtless a certain amount of curiosity must often arrest the attention of animals on perceiving the snake for two or three seconds before they become aware of their danger ; but of fascination as it is called, there appears to be none."

X.

BY THE SIDE OF AN AQUARIUM.

Are these the dirty mud-begrimed little fishes that were rescued from the slush and ooze of the paddy-field a fortnight ago? They look so wonderfully changed in appearance and manner. Instead of being ugly and dull, they are now very pretty and lively. What can this rainbow-coloured and fan-tailed fish be? Is it the *Khalisa* (*Trichogaster lalius*, Day), so common, yet so beautiful! Its body is vertically banded with scarlet and light blue, half of each scale being of one, and half of another, colour. The fins of the back and tail are dotted with scarlet. Every river and stream in Bengal is full of them; nor are they uncommon in the North West Provinces, and the Punjab.

A pretty little creature is that beautiful fish with vivid golden colour and black shoulder spots. Look how persistently it annoys and bullies the others; it is a little fiend. But we need not be surprised at its rude manners; a *tengrah* is always ill-tempered and unmannerly, and it was a mistake to place it with the others. The *tengrah* (*Macrones vittatus*, Block) of zoologists is very common all over India, Burmah, and Ceylon, inhabiting rivers, streams, and water-courses. It is a coarse and hardy fish attaining to seven or eight inches in length. Those found in dirty ponds and ditches are generally darker than those found in rivers, but all are liable to change colour. With regard to the habits of this species, Sir Francis Day, a distinguished naturalist, who has studied carefully the subject of Indian fishes and fisheries, records the following observation :—"This fish is termed 'the fiddler' in Mysore ; I touched one which was on the wet ground, at which it appeared to become very irate, erecting its dorsal fin and making a noise resembling the buzzing of a bee, evidently a sign of anger. When I put some small carp into an aquarium containing one of these fishes, it rushed at the small example, seized it by the middle of its back, and shook it like a dog killing a rat ; at this time, the barbles of the *Macrones* were stiffened out laterally like a cat's whiskers." Younger and smaller specimens may live amicably with other species.

Closely allied to the above species is the yellow *tengrah* found in Northern India, the Punjab and Assam. It is a smaller fish seldom exceeding four inches in length, is of a brilliant yellow colour, with black shoulder spots and four or five longitudinal lines.

Look at the pretty gambols of these tiny homely creatures. Active and inquisitive, some of them are busy exploring every part of the aquarium, while others have taken possession of a corner behind a small pot of plant, and are ready to fight in defence of their home against all intruders. One lay in ambush and has just surprised a *khalisa*. What are they ? Good heavens ; all this ado about some despicable *titpunti*. They are, no doubt, commonplace things, but at the same time they are objects of great interest to a student of Nature. The *titpunti* (*Barbus tetraurpagus, McClell*) and its allied species are very abundant in rivers, tanks, *jheels*, and water-courses in every part of India. They belong to the carp family, and enter largely into the diet of the poorer classes of the population. They are rather delicate fish and are generally preyed upon by the comparatively larger kind.

So, after all, the larger *magurs* have not agreed with the rest of the fishes, and are therefore kept in a separate tub with a number of *singhees!* The latter are very gregarious, and look like a thick black mass, as they lie hugging the bottom of the vessel. The *Magur* (*Clarius magur, Ham. Buch.*) and the *Singhee* (*Saccobranchus fossilis*, Day) belong to the scaleless family of Indian fishes having a smooth skin destitute of scales. They are found in ponds and ditches, and prefer muddy to clear water. Owing to the peculiar construction of their respiratory apparatus, they are enabled to carry on respiration for hours, and even days, without being in water. But we must observe it for ourselves, and ascertain how long they can remain alive, out of water or buried in the mud.

FINIS.

www.ingramcontent.com/pod-product-compliance
Lightning Source LLC
Chambersburg PA
CBHW032154160426
43197CB00008B/913